BERKSHIRE PATRIOTS

STORIES OF SACRIFICE

Dennis G. Pregent

Berkshire Patriots: Stories of Sacrifice

Printed in the United States of America
ISBN 978-1-956543-27-8

Book Design by CSinclaire Write-Design
Cover by Klevur

Testimonials

"What is it about the Berkshires that raised up so many honorable men and women, patriots who gave selflessly for America? Dennis Pregent has put his heart and soul into resurrecting the stories of twenty-eight citizens who achieved remarkable feats during wartime. His new book honors the sacrifices of these veterans from across four centuries: Soldiers, Sailors, Airmen, Marines—Patriots all."

— Admiral James Stavridis, USN (Ret),
served as the 16th Supreme Allied Commander of NATO;
author of **To Risk It All: Nine Conflicts and the Crucible of Decision**

"This monumental book based on original research and personal interviews is an achievement in perspective. His vivid tributes bring past and present inhabitants of northwestern Massachusetts into our awareness and admiration for the first time. From the French and Indian War and American Revolution to Vietnam and Afghanistan, these patriots have unstintingly paid the price to serve our country.

A veteran as well as a Berkshire native, Pregent places each of his heroes in a social and professional as well as a military context. Built one solid chapter at a time, *Berkshire Patriots: Stories of Sacrifice* is a new monument on the village green."

—Jim Clark, Ph.D., Professor Emeritus, NC State University

"Once again, Dennis Pregent has turned to Berkshire County, in this case, the small towns of Adams, North Adams, and Williamstown, Massachusetts, to give us a stirring picture of extraordinary things done by seemingly ordinary people. The men (and one remarkable woman) whose lives unfold in his pages served with valor, distinction, and often with great sacrifice.

His beautifully written book is a much-needed antidote to the cynicism too often present in our current circumstances. We come away from these portraits in courage with a renewed faith in the promise of America, with a conviction that, in times of national emergency, it is our best—our young men and women—who will rise to the occasion."

— Charles B. Dew,
Ephraim Williams Professor of American History Emeritus,
Williams College

CONTENTS

Williamstown, Massachusetts

Acknowledgments

AS ALWAYS, I AM most indebted to Carol, my caring, thoughtful wife and partner for fifty-seven years, who forever understands my imperfections and still tolerates, supports, and loves me after all these decades.

My greatest sources of information have been members of the Adams, North Adams, and Williamstown historical societies, many of whom I call friends.

Justyna and Gene Carlson from North Adams were an absolute font of knowledge, always on the lookout for news articles that would help me or the right book with some necessary information. Justyna passed away just before this book's publication; it is a huge loss to North Adams, and to me.

Paul Marino, another North Adams resident, met with me several times. He has his own historical website and freely offers information from his deep reservoir of North Adams history.

Eugene Michalenko, an Adams native, always took the time to lend me a book, email me an article, or point me in the right direction for some information I needed. He is Adams' walking history book.

Mike Kennedy from Williamstown, a former veterans' agent and now a friend, was a great help connecting me with veterans and their families. His devotion to veterans is beyond admirable.

I met several local historians who were helpful, including John Bordeau in Adams and Tom Boudreau in Readsboro, VT. Both enjoy and are willing to share their accumulated information on local veterans.

I had the honor to meet with Ed Marino and a volunteer group from North Adams who quietly go about restoring veterans' grave markers throughout the city, with no fanfare. They just do it out of appreciation for the men and women's service to their country.

Stanley Brown from Florida Mountain passed away during the writing of this book. He was a friend and a great source of local historical information.

I belong to a Senior Writers and Guests (SWAG) group, which over the years has provided much encouragement and support to my writings.

While the creation of this book begins with the author, I would not have known where to begin without Jane Hobson Snyder. She has been a wonderful help as a developmental editor. Every time I needed an answer to a question, she was there. Her editing skills are unsurpassed. Jane made my writing come alive, what more can be said?

Then, Lee Heinrich and her staff from WriteWay Publishing have been unsurpassed in helping me get this book across the finish line. We established a timeline, held to it, and the group managed the process (including me) so very effectively.

Also, a grateful thank you to the families who provided support, answered questions, responded to numerous queries, and supplied pictures of their loved ones, including John Bordeau in Adams and Tom Boudreau in Readsboro, Vermont.

A special thanks to Admiral James Stavridis, USN (Ret), who served as the 16th Supreme Allied Commander of NATO; James Clark, Professor Emeritus of English at NCSU; and Charles Dew, Ephraim Williams Professor of American History Emeritus at Williams College, for providing resounding endorsements of *Berkshire Patriots*. Tributes from these highly accomplished individuals mean so much and are greatly appreciated.

To anyone I may have missed: Thank You.

Honor Roll
Saint Joseph's High School Class of 1965

Many of you know that this journey began with my first book, The Boys of St. Joe's '65 in the Vietnam War, *inspired by a group of my classmates who, like me, joined up to fight and found ourselves in a distant land. Since that book was published, more information has surfaced about other classmates, and I'm proud now to be able to share a comprehensive list of these very young men and women who stepped out the doors of an idyllic North Adams high school and into the midst of international conflict and strife.*

Those Who Served

Michael Beaudin, Arthur Bullett, William Buzzell, Michael Chalifoux, Leo Chaput, Kevin Cummings, Joseph Daigneault, Gary DeMastrie, Norman Edgerton, Thomas Fallon, Russell Fitzgerald, Elizabeth Flick (O'Brien), Stephen Gilooly, William Gleba, Ernest Godbout, Michael Gorman, Thomas Josefiak, Nancy Kawa, John Kochanski, Edward Labonte, Edward Landry, Paul Langlois, Charles Lewitt, James Luczynski, Patrick Lupo, James Malloy, John Mangano, Walter Mille, Charles Moran, Kenneth Morgan, Graham Muldowney, Michael Ouellette, John Palmeri, Kenneth Phillips, James Quinton, John Raby, Ronald Racine, Peter Ronan, Russell Roulier, George Rusek, Leo Senecal, Barry Sullivan, and Bruce Witherell.

INTRODUCTION

Patriotism is a fundamental quality of character. . . . Character is often expressed in sacrifice; the willingness to give unstintingly of oneself has often been described as the "price we pay to serve."

—Admiral James Stavridis, in *Sailing True North*

WHAT IS IT ABOUT the Berkshires that inspired so many of its citizens to rise up, to represent their region in battle, to help form the nation they imagined it to be? In the beginning, it could have been simple geographic fortune: Massachusetts was an early colony, and the Berkshire Mountains in the far west of the state featured a promising intersection of textiles, education, and agrarian life. And again and again, over many wars and conflicts, heroes from the Berkshires—*patriots* in the best sense of the word—stepped forward and served.

Berkshire Patriots: Stories of Sacrifice offers a journey across centuries, with many chapters beginning idyllically in Adams, North Adams, or Williamstown, but then taking a sharp detour to battle-scarred lands or across distant waters. My goal in researching these local veterans, from the French and Indian War through the War in Afghanistan, was to enthrall, inspire, and captivate readers. Some of the names everyone will surely recognize, and others will be new, but by the end all should be unforgettable. Selflessness and courage abound, and bravery is commonplace. In some cases, peril became routine, and saving friends at the risk of one's own life a regular occurrence.

If I've done my job as a researcher and narrator, you will find yourself sharing these inspiring stories of honor and sacrifice with friends, coworkers, neighbors, and family. Some chapters will surely prompt smiles, others may yield disbelief, and many will bring sadness. Heroism is boundless and found in a broad cross section of society; patriots include men and women, enlisted and commissioned, bearing military ranks from private to admiral. For ease of reference, I've categorized the book into the three major centers of Adams, North Adams, and Williamstown. All subjects were associated with one of them, whether by birth, schooling, or retirement, though some were also connected to unincorporated towns or nearby areas like Cheshire, New Providence, and Pittsfield.

These twenty-five chapters cover a small sampling of people from the Berkshires who have sacrificed and served their country. *Stories of Sacrifice* begins with the French and Indian War, when Elisha Nims fights at Fort Massachusetts in 1745, and when Ephraim Williams, Jr., the founder of Williamstown and Williams College, is cut down in battle in New York State, leading his men against a superior force.

Chapters move through the Revolutionary War, Civil War, Spanish-American War, World Wars I and II, and on to Korea. Some served in multiple wars: Gordon Greene, North Adams' first casualty of the Korean Conflict, was a decorated soldier who had also served in the Buffalo Division during World War II, the only African American unit to see combat.

There's Jack Quinn from Williamstown, a World War II B-17 Tail Gunner who despite being repeatedly shot down over Europe, returned to fly in his deadly gunner's position until he was downed and finally wounded over Belgium. Again, he went back up until he achieved the required thirty-five missions and was relieved from duty.

More recent patriots include Peter W. Foote from North Adams, the decorated Drury High School athlete who served heroically in Vietnam and is a familiar name to many who still live in the area; and Michael DeMarsico, a North Adams hero deployed to one of the most dangerous areas in Afghanistan.

Dan Petithory from Cheshire was one of the first Green Berets to secretly enter Afghanistan after 9/11, working with a militia leader named Hamid Karzai, the future President of Afghanistan. Dan, with only ten other Berets and a small rag-tag local militia, went on to save a small town from 500 murderous al-Qaeda fighters.

One of the earliest stories herein features Joab Stafford, an early settler, who gathered colonists from an unincorporated area known as New Providence (which became Adams in 1780) and marched to fight in the Battle of Bennington in 1777. His group's bravery helped carry the day and many say affected the outcome of the Revolutionary War.

A little-known Medal of Honor winner from World War I, Charles Whittlesey, attended Williams College, then saved the famous "Lost Battalion" from total destruction. After the war, Charles disappeared mysteriously from an ocean liner.

Rudy Konieczny, a nationally known competitive skier from Adams, died on Mount Serra in World War II, killed when protecting his unit in the elite 10th Mountain Division. When found, he was said to be surrounded by eight dead enemy soldiers.

Some may have seen a recent PBS Special, "The Ritchie Boys," about a small secretive group of intelligence agents trained in Maryland during World War II. Michael Scarpitto, a thirty-three-year-old teacher from Drury High School, was in this select group and went on to serve heroically in Europe.

I had the honor of meeting one of the book's few living heroes, Marc Jaffe, a former Marine 1st Lieutenant living in Williamstown, who survived both decimating battles of Peleliu and Okinawa in World War II then, postwar, became the president of Bantam Books. His story stands out. Jaffe has lived several exciting lives, and (as he tells it) he is now working on his "second century."

Readers will thrill to the story of Ruth Koczela, who married a boy from Adams, attended North Adams State, and became one of the Navy's famous Code Girls in World War II. Under the supervision of Ruth and her fellow codebreakers, the Japanese Navy codes were unlocked, and the war shortened—some say by years—saving thousands of lives.

How did I gather such an eclectic collection of names? I was fortunate to be aided by dedicated professional and amateur historians from each town, a deeply caring former veterans agent from Williamstown who has become a friend, and families who reached out when they heard of my endeavor.

Writing about these patriots has been inspirational for me. I have enjoyed the historical research, meeting family members and descendants, and, in a few remarkable cases, talking to surviving participants. Their stories provoke a deep sense of admiration and appreciation of their unquestioned self-sacrifice, sense of duty, and integrity. All responded without question to their country in a time of need and were selfless in giving of themselves, some to the point of making the ultimate sacrifice.

This is my third book on the Berkshires. I've again chosen to select Adams, North Adams, and Williamstown, the three northernmost areas of the Berkshires, for the setting of *Berkshire Patriots: Stories of Sacrifice*. I know and love the region of my birth. As I write this Introduction on Veterans Day, I reflect on how my research allowed me to live in a bubble of patriotism and selflessness for these many months, collecting and crafting wonderful stories.

With the small towns and the City of North Adams so close together, I have had the opportunity to visit gravesites of many of these Berkshire patriots. The more I wrote, the more the visits became individual pilgrimages. The most poignant, for me, was visiting Dan Petithory's grave in the small town of Cheshire, in the company of his mom Barbara, his dad Lou, and brother Michael. Melancholy wrapped around us in the early afternoon visit not far from Dan's home . . . even twenty years after his passing.

Many of the patriots were buried under challenging circumstances. Elisha Nims' burial took place 255 years after his death. Some were interred not long after their deaths; others required families to wait for weeks or years for their loved ones' remains to be returned from Korea or Europe. In Sterling (Curly) Burnette's case, it was not until May 14, 1946, that the War Department notified his wife of his burial location: "The War Department is most desirous that you be furnished the burial location of your husband, the late Lieutenant Colonel Sterling S. Burnette, A.S.N. 0-383 615. . . . His remains are interred in the US Military Cemetery, Limey, France, plot H, row 5, grave 103."

Dick Burnette visits his brother Curly's grave in France. *[Courtesy of the Tarsa/Burnette family]*

I never did locate Ishmael Titus's grave. Ishmael was a former slave who was said to have fought at times shoeless in some of the most central battles of the French and Indian War and the Revolutionary War. He lived in Williamstown for well over the last twenty-five years of his life. I imagine his resting place has been lost to time.

Each of the cemetery visits had an effect on me, whether the monument was twenty years old or two hundred, smooth and bright or chipped and weathered. Most times, I found myself thinking the setting was too simple or humble for what had been accomplished by the person resting within. Near the end of this book, I've included a sampling of some "pilgrimage" sites that I found meaningful, locations that bring local history into particular focus. History is all around us: Someplace you drive by every day might suddenly feel magical after you realize its meaning.

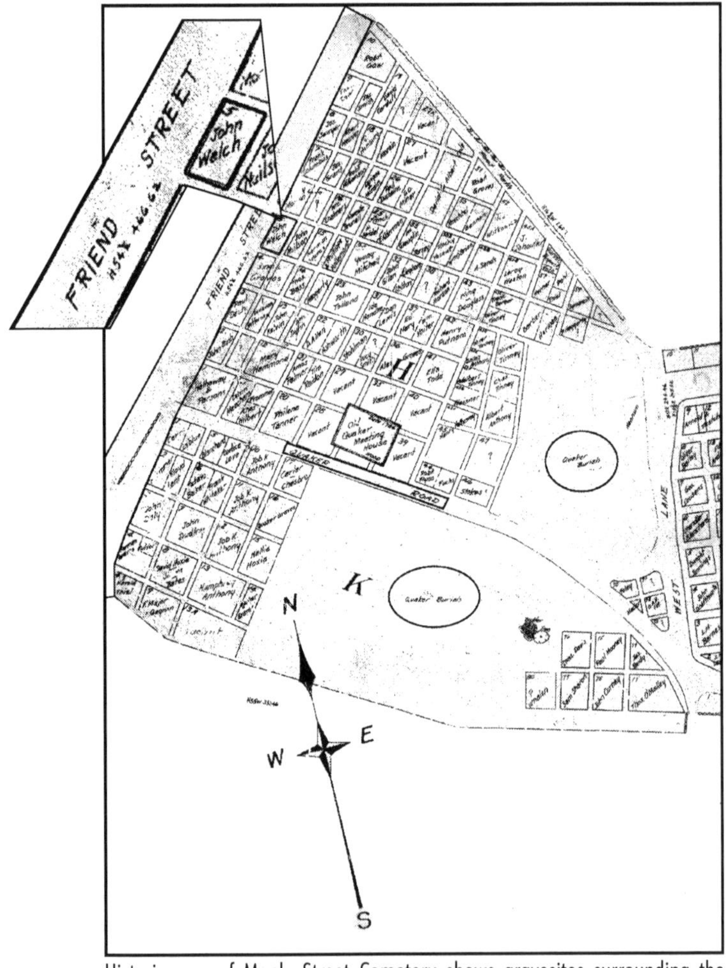

Historic map of Maple Street Cemetery shows gravesites surrounding the Quaker Meetinghouse, including that of John Welch, a Civil War quartermaster. *[Courtesy of the Town of Adams]*

Originally, I thought there might not be enough tales to warrant a complete book, but I have come to realize far too many stories deserve to be told. Consider Sergeant Richard E. Welch, a Civil War Medal of Honor winner who lies in near-anonymity in Williamstown's Eastlawn Cemetery just feet off the main street. Or Ensign Bernard St. John from Adams, a pilot on the carrier *Intrepid*, winner of the Navy Cross, who helped sink a Japanese battleship; or Jeremiah Colgrove, a Revolutionary War sailor and an early settler of North Adams, who once owned and was buried at Colgrove Park, now interred with relatives at Southview Cemetery. Many future stories beg to be told . . . .

It's likely none of the subjects featured in *Berkshire Patriots* ever considered themselves heroes, but long after they're gone, that's what they have been to me. Their strong sense of duty, perseverance, selflessness, and commitment inspire my deep admiration and wish for emulation. Their stories continue to uplift my soul and refresh my thoughts about the goodness of mankind.

Novelist Bernard Malamud said it well: *Without heroes we are all plain people and don't know how far we can go.*

Now I leave you to meet these remarkable *Berkshire Patriots.*

Dennis Pregent, Marine, Vietnam Veteran
November 2022

BERKSHIRE PATRIOTS

STORIES OF SACRIFICE

Adams/Cheshire, Massachusetts

Colonel Joab Stafford
Battle of Bennington
REVOLUTIONARY WAR

It was August of 1777. Bands of soldiers and militia from New Hampshire, Vermont, Massachusetts, and New York, answering an urgent call for help, undertook a forced march to a small hill outside of Bennington, Vermont. A forty-eight-year-old captain, Joab Stafford, while leading a mixed band of militiamen, attacked uphill into the center of a defensive British breastworks through a maelstrom of gunfire. In hand-to-hand combat, he and his men routed a large force of Tories and German Hessians. The group's bravery carried the day, and many say the Battle of Bennington affected the outcome of the Revolutionary War.

The Revolutionary War had begun at Concord and Lexington, Massachusetts, in April 1775 and would last more than eight years. The war's peace treaty was signed with Britain in August 1783.

Earlier Background

THOMAS STAFFORD, REPORTEDLY THE first Stafford in America, came from England in the 1600s. Little is known of his life, although he is credited with bringing the family's coat of arms with him. It was engraved on a foot-square wooden panel with the words "Virtue—the Corner Stone of Life."

Samuel Stafford, son of Thomas and grandfather of Joab, married Mercy Wescott. It's said that the Wescotts, persecuted for their beliefs, were driven from Salem, Massachusetts, along with Roger

Williams, a puritan minister, and ultimately established and settled Providence, Rhode Island.

Samuel and Mercy's son, Thomas, married Audrey Green. She was his second wife and together they raised fourteen children, one of whom was Joab. Records indicate Joab was born on November 14, 1729, in Coventry, Kent County, Rhode Island. On October 6, 1751, at the age of twenty-one, Joab married Susan Spencer in Greenwich, Rhode Island. He and Susan, described as a pretty Quaker girl, raised ten children together.

Joab became a successful lumber merchant, a landholder, and travelled to other countries buying exotic wares to sell in the Americas. It is said he was elected to the provincial assembly in Coventry and also became involved in its militia. Records indicate he was referred to as Captain; it's uncertain whether this title was related to his militia service or to his naval experiences.

In 1767, at thirty-eight years old, Joab was hired to survey land in the Berkshires. Attracted to the area, he bought 400 acres, and joined by other families from Rhode Island, he moved his family there. The early settlers called their new home New Providence, now known as Stafford Hill in Cheshire, Massachusetts. It was more than 150 miles inland from Joab's birthplace in Rhode Island.

With the establishment of tradesmen's shops, taverns, and a stagecoach route, Stafford Hill became the center of commerce for the rural area. Many of the new settlers were Quakers and Baptists, and often patriots and political activists. Once settled there, they erected the first Baptist church west of Connecticut in 1774. The Stafford home, tavern, and store became a local gathering spot for the growing citizenry.

By the year 1777, the Revolutionary War had been in process for several years, with battles and skirmishes won and lost by the colonists. As the war see-sawed, British Major General John Burgoyne (known affectionately by his soldiers as "Gentleman Johnny"), proposed a plan to split New England from the other colonies. His plan would isolate the northern colonies, denying them supplies from the middle states.

The British Plan

General Burgoyne proposed leading 8000 troops from Canada south toward Albany, New York, via Lake George. While this was occurring, a diversionary force of 1200 led by General St. Leger would sweep across Lake Ontario to Oswego, New York, marching east and proceeding down the Mohawk River. St. Leger and Burgoyne would join their superior, General Howe, in Albany.

With King George III's approval, each group began to amass supplies and men. Burgoyne's troops included British and Canadian soldiers, German Hessians (mercenaries), and a number of Indigenous tribesmen. There were dismounted dragoons, infantry, marksmen accompanied by artillerymen, the regimental band, women and children, and other camp followers. The General's supply chain was of concern to some from the beginning of this operation. Not only did it extend from the Canadian frontier to Albany, but the goods themselves came by sea from England, a notoriously long, arduous, and undependable process.

The light infantry used in reconnaissance were known as *rangers*, quickly deployed and fast afoot in rough terrain. *Dismounted dragoons* rode into battle on horses but fought on foot; therefore, they were not as easy to deploy as unmounted rangers. At one point, it was suggested to General Burgoyne that he use the rapidly deployable rangers instead of dismounted dragoons. He spurned the suggestion, believing supplies of rations, horses, and wagons would be easily gathered along the march, either forcefully taken from the rebels/colonists or provided/bought from the Loyalists. (It is now estimated that colonists loyal to the Crown represented one third of the population in the American States, though Burgoyne may have overestimated that figure in his time.)

Burgoyne's confidence never waned, and the invasion force of approximately 10,000 souls departed Canada mid-June 1777. The procession stretched for miles, using two-wheeled carts to transport supplies, ammunition, artillery, and personal items. Within this group, the 7000–man fighting force was composed of approximately half British soldiers and half non-English speaking German mercenaries (sometimes known pejoratively as "brunswickers" by the colonists).

By early August, General Burgoyne's force (still some miles from Albany) was in woeful need of transportation. He left Canada with fewer than half the carts he needed, and only a third of the horses for his troops. Most of his dismounted dragoons, lacking horses, were trudging through the wilderness in their heavy uniforms, large pointed hats, and unwieldly sabers, drastically slowing the column's progress.

As the expedition proceeded, it had early successes, including the capture of Fort Ticonderoga, although supplies became scarcer. Hearing from a Loyalist spy that nearby Vermont towns had generous amounts of wagons, livestock, and ammunition/powder, General Burgoyne commissioned Lieutenant Colonel Baum, a German officer who did not speak English, to lead a strike force charged with securing and capturing provisions, expecting the generous help of local Loyalists.

Burgoyne gave Baum 760 men, including soldiers, dragoons, Loyalists, marksmen, and artillerymen with three-pound cannons. One hundred and fifty Native Americans accompanied the group with the purpose of conducting predatory raids ahead of the expedition, alarming residents and any who might be encouraged to help the colonists. Baum also had about 300 Loyalists (Tories) in his group, totaling about 1200 fighters.

As the mission proceeded, Loyalist reports indicated that substantial provisions and livestock were available in Bennington, in present-day Vermont. Lt. Col. Baum was directed to that area. (At that time, newly independent Vermont was still referred to as Hampshire Grants.)

On their route, Baum's troops had ongoing skirmishes with colonists and, approximately five miles outside of Bennington, he realized he was facing a larger force than expected. Baum stopped, requested help from General Burgoyne, and set up several defensive redoubts, constructed of earth and timber on a nearby 300-foot hill overlooking the Walloomsac River. He then placed troops on both sides of the bridge that crosses the river, leading directly into Bennington.

The colonists, aware of Burgoyne's expedition and its successes, became alarmed and sent out calls for help to militias in Vermont, Massachusetts, New Hampshire, and New York. The borders of Massachusetts and New York are fewer than ten miles from Bennington; New Hampshire is fewer than fifty. The states quickly reviewed their "Alarm Lists," rosters of able-bodied men, and "Train Band Lists," rosters of those with some military training, and called local gatherings in preparation to march to Bennington.

With Native Americans sighted and rumors of the advancing British army, colonists and their families were in a frenzy. Tories' properties and goods were seized, sold, and the monies used for purchasing supplies. Families helped by melting their pewter and molding musket balls. They also prepared themselves to flee. Even church parsons were involved, making fiery self-defense sermons and often secretly concealing gunpowder under their pulpits.

The alarm reached New Providence, and three companies were formed from four townships: New Providence, Lanesborough, East Hoosick, and Gageborough (Windsor). It was determined that Captains Stafford, Lowe, and Brown would each lead about forty troops, trained on the level spaces atop Stafford Hill.

The groups gathered at Joab Stafford's house on August 12, 1777, and made hasty plans to help the Vermonters.

Unofficially, the groups called themselves the Silver Grays. The name is presumed to be related to their relatives who served during the French and Indian War. A Silver Gray Muster Roll lists over 100 names and petitions for the commissioning of their officers. They all signed the papers of enlistment in August 1777.

On August 14, after several days of preparation, the groups began their thirty-plus-mile forced march to Bennington, collecting additional fighters along the way. The march continued on the next day through torrents of rain, which by great fortune had delayed any military action in Bennington.

In the early summer, as the British had advanced, Colonel John Stark, a former Roger's Ranger and a veteran of the battles of Bunker Hill, Canada, and Trenton, had assembled three New Hampshire regiments; in July he had begun organizing the groups. He had coordinated his efforts with other militia leaders (Seth Warner and William Williams) and the Continental Army's General Benjamin Lincoln. At this point, approaching Bennington on a mid-August day, it was agreed that Stark possessed the most experience, and the collective fighters selected him to lead the entire group in defense of Bennington.

Six Aged Citizens of Bennington, 1848. Sons of Ethan Allen's Green Mountain Boys. All but one fought beside their fathers at the Battle of Bennington. *[Original daguerreotype by Charles H. Gay and enhanced by Kingston Studios ca. 1900. Courtesy Bennington Museum Collection, Bennington, Vermont, gift of Mrs. Robert M. Parmelee, catalog number 1966.1875.]*

The deluge on August 15 allowed Colonel Stark more time to establish his tactical plan. Stark found himself with 900 New

Hampshire militia, 200 Vermont militia, 200 Continentals, 200 Green Mountain Boys, roughly 280 Berkshire men, and 100 Stockbridge Indians—somewhere between 1500 and 2000 men. It is said that almost every able-bodied militiaman from nearby Williamstown came to help.

On the 16th day of August, the rain ceased, and without a cloud in the sky, late-summer mugginess crept in. It had no effect on the colonists who were wearing light shirts and pants, but on the invaders and their warm wool uniforms, it took a toll.

The Battle

On the morning of August 16, 1777, Colonel Stark deployed groups of men to envelop the British breastworks (also known as the dragoon and Tory redoubts) from the right and left sides. Attack on the smaller Tory breastworks was assigned to the Silver Grays and other units. Colonel Stark himself would lead his New Hampshire militia and attack Lt. Col. Baum's main position near the bridge. The men were advised to stay hidden until they heard two musket shots to commence their assaults.

Colonel Werner told Captain Stafford that his group of volunteers would assault the Tory redoubt. Aided by local colonists, they traveled toward the earthworks in a hidden ravine that brought the men just below the redoubt. Tories spotted his group, right under their guns, and shot downward over the breastworks. Discovered, Captain Stafford, leading his men, rushed up the hill to breach the redoubt. Almost immediately, Stafford was shot in the foot ("cutting some sinews") and fell to the ground. Quickly recovering, he sprung forward, encouraging his men to clamber over the breastworks while the Tories were reloading.

As the volunteers breached the breastworks, the Tories and cannoneers panicked, abandoned their redoubt, and ran down the hill, seeking safety with other British troops. The colonists pursued them and many Tories were killed, wounded, or captured. Simultaneously, the dragoons were routed from their redoubt; in the battle, Lt. Colonel Baum was mortally wounded.

The fighting, which commenced around 3:00 p.m., ended several hours later with a complete victory for the colonists. Fatigued by their efforts, they faced another challenge when 650 British reinforcements sent by General Burgoyne and led by Lt. Colonel Breymann arrived. Fortunately for the colonists, Major Warner also arrived, accompanied by 150 fresh militia troops from Massachusetts.

The fighting was heated and often hand-to-hand. The Americans were tired from constantly firing at the British. Sometimes their Charleville muskets would become too hot to handle, and they would grab a Brown Bess from a dead Hessian to continue the fusillade. Another time, the colonists were able to capture a British cannon and use it against the British. The colonists outflanked the British position, forcing their retreat in a disorderly fashion. Fighting ceased at darkness.

The morning light revealed a resounding victory for the colonists. Colonel Stark's troops had killed over 200 Hessian mercenaries, wounded many others, and captured over 700 enemy troops. Stark's losses were 30 colonists killed and 40 wounded. Four of the dead were from Bennington and returned to their families for burial.

On this hot, damp August day, a local farmer named Mister Harmon dragged more than 160 of the enemy dead into two large excavations normally used to store potatoes in the winter. This burial area was the final resting place for most of the invaders.

Carts hauled away the wounded. The prisoners included 30 officers, hundreds of Hessians, and 155 Tories. The colonists captured four cannons along with their ammunition carts, twelve drums, 250 sabers, and hundreds of arms.

The prisoners were closely guarded and marched through the streets of Bennington. Tories suffered an additional indignity, being bound together in twos. Temporarily, the captured troops were packed into the town's meeting house, eventually to be marched to

Boston and repatriated. Captured British officers spoke of colonists as the "enemy," but when speaking about Tories or those loyal to the Crown, they referred to them as "damn traitors or scoundrels." Even the British held the Tories in low esteem: apparently, no one likes a traitor, even the one helping you.

The colonists' overwhelming victory sent encouragement throughout the colonies, prompting the French to take a more supportive role; discouraging many Loyalists, Canadians, and Indigenous tribes who were assisting the Crown; and strengthening the opponents to policies of King George III. Beyond the battlefield, the colonists had flexed their might simply by denying supplies, horses, wagons, and other livestock to General Burgoyne's invading forces.

The battle destroyed almost 15% of Burgoyne's army, increasing the hope and confidence of American patriots. Many historians believe Burgoyne's loss at Bennington led to his eventual defeat several months later at Saratoga, helping to ensure the success of the American Revolution.

As a postscript to Colonel Stark's significant victory, he offered a $20 reward to anyone who helped him find the thief that stole his brown mare. She was described as five years old with a star on her forehead. The mare had been taken on August 16 right in the middle of all the fighting and was never recovered.

Notable Participants

One militia participant and New Providence citizen was Daniel Read, maternal grandfather of Susan B. Anthony, the renowned women's suffragist. Read was a Quaker who had temporarily set aside his peaceful beliefs "in defense of their homes and liberties" (*Adams Historical Society Newsletter*). He fought by Stafford's side in the Battle of Bennington and was among the troops who fought in Québec with Benedict Arnold and at Fort Ticonderoga with Ethan Allen.

Later in life, Read participated in Shays Rebellion, an armed and violent protest against foreclosing on farms of veterans of the Revolution. Six hundred participants from the Berkshires attacked the federal arsenal in Springfield. They were repelled, and eventually federal troops came to Stafford Hill and arrested Read and other participants. They were not jailed but were required to swear an oath of loyalty to a court official.

Another participant of note was Ebenezer Webster, a militiaman, innkeeper, and later a judge. He was the father of Daniel Webster, orator, lawyer, and future member of Congress.

Uniforms

Much has been written of British soldiers' extravagant uniforms of the era, and their strategic benefit or detriment depending on the moment. Long wool red coats were distinctive paired with off-white breeches, white vests, black garters, black boots, and a trimmed black cocked hat. Different rank insignias finished the look. Officers wore sashes, shoulder belts, and swords. Hessian allies wore blue coats made of coarse material like burlap, black boots, and tricorne hats, with grenadiers wearing tall hats. Hessians carried slender swords.

American soldiers prior to 1779 wore a wide variety of modest long brown coats. In 1779, Washington ordered blue coats with white waistcoats. Many states then added different linings, facings, and buttons. Patriots also wore off-white or beige waistcoats, breeches, long sleeve hunting shirts, black tricorne hats, white stockings, and dark shoes with buckles. At Bennington, because it was a hot sultry day, and because the call went out for rapid assembly, most colonists wore just hunting shirts and pants.

Weapons

Most English soldiers used a Brown Bess musket. The smoothbore, muzzle loading, flintlock weapon fired .75-caliber one-ounce musket balls. The average soldier was capable of firing two or three rounds a minute, and accuracy was limited to 80 yards. The Hessians used a Potzdam musket, manufactured in Germany. It was .71-caliber and sported an 18 inch triangular bayonet.

The colonists used a variety of muskets but most common was the imported .69-caliber Charleville provided by France. The Charleville had a 20-inch overall socket bayonet. They were also known to use the Brown Bess, and a few men had long rifles they could fire effectively up to several hundred yards, although these rifles had no bayonet lug. The Brown Bess, Charleville, and Potzdam muskets all weighed around 10 pounds, notable compared to a modern-day weapon such as the M-16, which weighs 7 pounds.

While the muskets were not effective much beyond 80 yards, a man, if hit by one of their musket balls, suffered a frequently lethal wound. Often when a soldier was hit in the chest, the doctor would just set him aside to expire. If a fighter was hit in the leg or arm bone, the lead ball would disintegrate the bone, requiring amputation without anesthetic.

The bayonet was the most frequently used edge weapon, often relied upon after the first volley while participants were reloading. The three-sided bayonet made nasty, deep wounds. One colonist leader was quoted as saying, "the musket [made] a good handle for the bayonet."

Other edged weapons such as swords were widely used by infantrymen. Officers usually had shorter swords, while cavalrymen had longer sabers. Smoothbore flintlock pistols were used by both sides with a limited range of twenty feet.

The English employed three or four smoothbore, muzzle-loading, three-pound cannons at Bennington. The poundage was determined by the weight of the shot being fired. The colonists used captured cannons against the British troops.

The Stafford Hill Memorial stands at attention across the Hoosac Valley from Mount Greylock. *[Nick Mantello]*

In the center of the memorial tower lies the final resting place of Joab Stafford. *[Nick Mantello]*

EPILOGUE

After the battle, Joab Stafford was carried by litter back to Stafford Hill to recover. Records indicate that in his later years he was categorized as an invalid and received a pension for his service.

Joab and several others petitioned the General Court of Massachusetts Bay on October 14, 1777, to officially recognize the independent company called Silver Grays and its commissioned officers (unofficially established at Colonel Stafford's home on August 12, 1777) and to "procure pay for services which we have already done," having distinguished themselves in the Battle of Bennington just months prior. It appears that sometime after the Battle of Bennington, Joab was promoted to the rank of Colonel.

In 1783, Colonel Stafford began to sell off his property and eventually moved to Albany, New York (40 miles further inland), becoming an assemblyman and a tavern owner. In 1795, his wife died, and Joab moved temporarily into the home of his son-in-law. Age and illness found him returning to Cheshire where he died on November 22, 1801, at the age of 72.

Joab was initially buried in old Cheshire Cemetery, but being a town hero, he was reinterred at the Stafford Monument on Memorial Day 1927. The grave monument was erected that same year by the Sons of the American Revolution. The memorial tower constructed of fieldstone is a replica of one of the earliest known structures in North America, a stone mill believed to be of Viking origin in the area of Newport, Rhode Island. Five hundred people gathered for the ceremony.

His tombstone reads:

In memory of Col.
Joab Stafford who
Fought and bled in his
Country's Cause at
The Battle of Bennington
August 16th, 1777, & who
departed this life Nov 22,
1801, aged 72 years
and descended to the tomb
with an unsullied reputation

Of the 600 Berkshire-area men who responded to the call at the Battle of Bennington, it's unclear precisely how many perished and where. Sketchy records indicate some were killed there, then returned home for burial in family cemetery plots. Others were interred in unmarked graves, the sites known only to God.

In 1889, a battle monument was erected in Bennington near the site where supplies were stored, those that Burgoyne desperately needed to capture. The obelisk, located in Bennington, is actually ten miles away from where the fighting occurred in Walloomsac, New York. The commanding obelisk is over 300 feet tall; from its 200-foot observation post, visitors can see the nearby states whose militiamen lent such critical support that August day in 1777.

Silver Grays Epilogue

After the battle, the Silver Grays returned to their farms, scattered, and after some time the name fell into disuse.

In August 1971, the Adams, Massachusetts, fire chief conceived the idea of reconstituting the "Grays" to participate in a reenactment ceremony celebrating the Battle of Bennington. In the summer of 1972, the group gathered, donning colonial attire and carrying period muskets, and marched the twenty-two miles to Bennington to celebrate the colonists' victory almost two centuries earlier.

The new "Grays" became popular and were requested for many years to perform in parades, festivals, reenactments, and celebrations on Stafford Hill.

Quartermaster John Welch
USS *Brooklyn*
Civil War

John Welch was on the quarterdeck of the sloop of war USS Brooklyn, *helping the helmsman navigate the narrow passage into Mobile Bay. The* Brooklyn *was under fire from the rebels at nearby Fort Morgan. Accompanied on its port side by another sloop, the* Octorara, *the two were part of an 18-ship flotilla intended, once and for all, to complete the blockade of the last important port of the Confederacy on the Gulf of Mexico: Mobile Bay.*

The cannonade from the fort was intense and devastating. The Brooklyn *and* Octorara *were the lead sloops, and once they ran the gauntlet of shelling, more danger was still to come. Awaiting them was a small, determined rebel fleet and its iron-clad monitor.*

LITTLE IS KNOWN OF John Welch's early life. He was born on June 15, 1829, in Schaghticoke, New York, to Mr. and Mrs. Robert Welch, recent Scottish immigrants. Schaghticoke was a small village with a population of around 1200 people, located thirty-three miles west of Williamstown, Massachusetts, where the Welch family moved when John was five years old.

Presumedly, John's dad was looking for work in the local grist-mills or sawmills, which drew recent immigrants from all over the region. The family settled in, enjoying the Berkshire landscape. John, as a boy, hunted and fished and taught himself to swim in the nearby Hoosic River, which ignited his interest in waterways for the rest of his life. Many believe that's where his seafaring story begins.

Early information describes John as no more than a "stripling" when he first found a place on a merchant vessel; likely, he was thirteen or fourteen years old.

As an adolescent, he found himself shipping out on different merchant ships traveling across the globe carrying supplies and trade goods to and from ports throughout the world. No matter the vessel, as an inexperienced youth, he would be assigned the most menial tasks while he learned the trade of a mariner. It was difficult and exhausting work: scrubbing decks, climbing masts, furling and unfurling sails, standing watch, and completing any task the captain ordered him to do.

As he became more experienced, he found himself preferring the clipper ships. The sleeker ships were an American innovation created during his time at sea that had longer profiles and more sails, enhancing a ship's speed. It allowed more rapid delivery of cargo, thereby releasing the mariners to go on liberty more frequently.

Three times, John's travels brought him off the coast of South Africa and around the Cape of Good Hope as his ships transported goods to the Far East. The older crew members surely regaled him with the story of the legendary *Flying Dutchman.* The *Dutchman* was said to be crewed by ghostly sailors doomed forever to struggle through the nearby waters, never able to round the Cape.

A much more dangerous setting, especially for nineteenth-century sailors like John, was sailing around Cape Horn in southern Chile. Its strong winds, large waves, and brisk current made it a natural graveyard for sailing vessels. John and his shipmates made this passage three times as well, under the most severe conditions, always watching for the additional hazard of icebergs.

It was said that a sailor who successfully rounded the Horn was entitled to wear a gold earring in his left ear, though it's unknown whether John accepted that privilege. Cape Horn had its own tales to tell. In 1788, an unsuccessful navigation became the stuff of legend, as portrayed in the Marlon Brando film *Mutiny on the Bounty,* based on a 1932 book of the same name. As recounted with some historical accuracy, the perilous Cape passage was the primary reason for the mutiny of its crew.

John's ship encountered no mutinies, but near-death adventure was never far away. Once, near the South China Sea, when his merchant vessel strayed too close to shore, it struck rocks and sank in the "chokepoint" area of the Malacca Strait, a treacherous body of water that separates Malaysia and Indonesia. The Strait is 550 miles long, and the chokepoint is less than 1.5 miles wide at its narrowest point. Over the centuries, it has been the source of many shipwrecks.

The entire crew managed to swim to shore in a desolate area where they remained for days scrounging for bits of food, killing and parceling up a few skinny seabirds, and drinking water from rocky holes. They were marooned for so many days that starvation seemed assured when, miraculously, one of their members attracted a passing vessel, and they were rescued.

Not long after this harrowing experience, in 1862/63 John decided to join the US Navy.

Several years before John's enlistment, in 1859, the Navy launched the USS *Brooklyn*, a wooden sloop of war commissioned with Captain David G. Farragut in command. (He would go on to become Admiral Farragut.) The sloop was named after the borough of Brooklyn. She was 233 feet long, 43 feet across, and had a 16-foot draft (the depth of water required to sail). It was the time in history when sailing ships were converting to steam; consequently, the *Brooklyn* was powered by a steam engine that used coal and had three full masts of sails (a hybrid).

She had a crew of 335 officers and sailors and a notable battery of one ten-inch and 20 nine-inch smoothbore guns. After her initial launch, she was active in patrolling the Caribbean, the coast of Mexico, and South America.

When war was declared in April 1861, the USS *Brooklyn* became actively engaged in protecting federal properties, forts, and installations. In 1861, at the war's outbreak, the *Brooklyn* tried but failed to assist in resupplying Fort Sumter.

Shortly after the war began, President Lincoln ordered the ports of all seceded states to be blocked; the fleet took on the considerable task of intercepting Confederate ships attempting to outrun the blockade, ships that became known as blockade runners.

Much of this action took place along the coast or in shallow harbors, necessitating smaller craft and more river gunboats. The Union navy, composed of fewer than 50 boats at the war's beginning, grew to over 600 by war's end. The need for sailors was drastic and immediate. Starting from a force of 9000 sailors, it grew to almost 60,000 by 1865.

Many recruits had little experience, although sailors from the merchant marine, like John, were welcomed, highly prized, and given advanced ratings. It's interesting to note that sailors were often characterized as independent souls. Unlike soldiers, they were not farmers and preferred adventure to stability.

As a qualified seaman with vast mariner experience, John was assigned immediately to learn the responsibilities of a US Navy Quartermaster. Many of these duties he had already experienced, such as standing watch on the "quarterdeck," serving as helmsman, helping with navigation, and other bridge watch duties.

In this junior petty officer position, he would be in one of the oldest rates in the Navy. John's domain was the quarterdeck, where the ship's steering wheel is located. He also used the ship's navigational aids and helped keep its charts up to date.

Over the years, John had a reputation of "excellent service in sounding to ascertain the depth of water." Whenever his ship would near land, especially in coastal waters, John would use a centuries-old navigational aid that allowed the ship's captain to know the depth, position, and type of seabed the vessel was passing over.

He would swing the leaded weight ahead of the ship. It then dropped into the water and sank to the bottom so that by the time the boat reached it, he could read the markings on the rope and call out the depth in "marks"—one mark equaled a fathom (six feet). The leaded weight had a secondary use; its hollow base picked up samples from the seabed, which also helped determine the ship's position.

Even in his position of authority, John bristled a little at his new disciplined life versus the one he had enjoyed for nineteen years on merchant vessels. He now wore a blue, wrinkled uniform with a round cap and had a defined schedule, compared to his former casual one. When the Marine bugler sounded reveille, John needed to roll out of his hammock right away and store it topside. Before breakfast or a smoke, the ship must be cleaned, the decks sanded and washed down. Then it was on to mess duty for breakfast, after which came more cleaning and polishing. Life was far more regimented on this sloop of war than it had been on various merchant ships.

After a while, John got used to the routine, and some of the more demanding tasks were completed by the landsmen (recruits with no experience who performed menial and unskilled tasks). He did enjoy his pint of coffee at breakfast and a bit of salt beef. Lunch was often more beef or some pork, vegetables, and coffee. There might be a light meal in the evening. Food was reliable.

Blockade duty was mostly monotonous, but they needed to be ready for the enterprising blockade runner. The gunner's mates were constantly cleaning and inspecting their guns, ensuring they were prepared for immediate use. Particular attention was paid to the guns later in the evening, for it was in the hours of darkness that blockade runners were most active.

During many tasks performed in the merchant service, the men might sing ditties. Not so in the Navy. There was silence, so one could hear the petty officer's orders or the boatswain's pipes. He did like playing dominoes with his mates, and although gambling was forbidden, John managed to know where to find an active dice game, and he often won.

John received petty officer pay of $20.00 a month and a daily ration of grog. Not long after he enlisted, Congress abolished the ration because of abuses, and the men received five additional cents a day in lieu of their spirit. The $1.50 monthly allotment did not stop sailors from using their ingenuity to smuggle liquor on board for the duration of the war.

Sleeping below deck could be oppressive. Hundreds of men occupied a small, barely ventilated space. Many tried congregating around the hatches for fresh air, although the Marines were ordered to disperse large groups.

For the blockaders, boredom was their primary enemy, and they often welcomed any distraction in their everyday routine. It could include taking on coal or supplies in a port or being in dry dock for repairs, both hopefully providing opportunity for shore liberty.

Tattooing was popular on board. Several of John's shipmates had tattoo kits; with a ready supply of needles and India ink, for the right price they would be very creative in applying "emblems" anywhere on the bodies of fellow crew members. Most popular were anchors, unclad mermaids, crossed cannons, "Mother" scrolls, daggers piercing a heart, and crosses. We know John had many of these memorialized on his body, including several works of art that include Admiral Farragut's likeness on his right breast and an American eagle across his back from shoulder to shoulder.

By late April 1862, John's USS *Brooklyn* had become part of Admiral Farragut's flotilla that was bombarding Fort St. Philip and Fort Jackson on the Mississippi River in a vital battle for possession of the City of New Orleans.

In the early hours of April 24, John helped guide the helmsman through a dismantled chain stretched across the river, attempting to slip by their guns in early-morning darkness. Discovered, the *Brooklyn* took fire from both forts; luckily, their aim was poor (as was *Brooklyn*'s), so only minor damage occurred as the ship continued upstream.

Once past the forts, the action heated up, and during the ensuing naval battle, the CSS *Manassas*, an iron-plated, turtle-looking monitor sitting low in the water, rammed the *Brooklyn*. Unbeknownst to her crew, it caused severe damage to *Brooklyn*'s hull. (Notably, a short time later, another Union ship poured cannon fire into the *Manassas*, and she eventually exploded.)

The fleet continued battling its way upriver, losing one vessel but sinking twelve of the Confederate ships. Arriving in New Orleans, Admiral Farragut demanded and accepted the city's surrender. The *Brooklyn*, realizing the need for repairs, required a 24-foot-long piece of planking patched over the tear in her hull.

Of greater loss, the *Brooklyn* also sustained casualties: eight men killed and twenty-one wounded. Senior Quartermaster James Buck was awarded the Medal of Honor during this action. His citation reads:

> *Although severely wounded by a heavy splinter, Buck continued to perform his duty until positively ordered below. Later stealing back to his Post, he steered the ship for eight hours despite his critical condition; his bravery was typical of the type which resulted in the taking of Forts Jackson and St. Philip and the capture of New Orleans.*

As part of the quartermaster group, John was in the midst of all the action, presumedly aiding Buck when needed—then assuming his duties as the senior quartermaster when Buck spent months recovering.

After capturing New Orleans, the Confederacy's largest city, the USS *Brooklyn* headed upriver and accepted the surrender of Baton Rouge, Louisiana, and Natchez, Mississippi, and then returned to New Orleans for reassignment.

Some historical accounts consider the fall of New Orleans to be more deeply influential than it may have seemed at first glance; its loss may have influenced European countries to decline to recognize the Confederacy diplomatically.

The USS *Brooklyn* continued chasing blockade runners for the next twenty-four months, with detours to the New York Navy Yard for repairs. The ship was assigned to blockade duty off Mobile Bay in October 1862, then headed to Galveston, Texas, where she captured a cotton-laden sloop, the *Blazer,* and another called the *Kate,* and aided in destroying other blockade runners. In July, she returned to New Orleans for blockade duty.

The *Brooklyn* was refitted/recommissioned in April 1864 and put back out to sea under the leadership of Captain Alden. The ship joined Admiral Farragut's squadron off Mobile Bay.

The upcoming Battle of Mobile Bay was important. It would determine control of the last major port of the fading Confederacy, and John Welch was to be front and center in the action. On August 5, 1864, Admiral Farragut mustered his squadron of 18 ships (including four iron monitors) just outside the bay.

Battle of Mobile Bay, painting by J.O. Davidson, 1886. The USS *Brooklyn* is on the far right, leading the outer line. Confederate ships are on the left. *[Courtesy of Adams Historical Society]*

Forts guarded the entrance, which had been narrowed by a minefield to create a perilous passage for Farragut's fleet. With mines scattered on the left and right of the entrance and Fort Morgan on the right, Farragut's four monitors led the way, which he hoped would absorb most of the fort's cannon fire. His lead monitor, the *Tecumseh*, almost immediately hit a mine (which was at that time called a *torpedo*) and was sunk.

This caused confusion, and the remaining three monitors were now in disarray.

Farragut's fourteen ships entered the bay in a column of twos behind the monitors and came under intense fire from nearby Fort Morgan. Hemmed in by the minefield and cannon fire, Admiral Farragut directed his flagship, the *Hartford*, to lead the main column of ships directly across the minefield.

And then he reportedly said, *Damn the torpedoes, full speed ahead.*

John, on the quarterdeck of the USS *Brooklyn*, was just to the left of the *Tecumseh* when she sank. The sailors were under concentrated fire from the fort, and the helmsman was concerned about some suspicious buoys in the water amidst the heavy smoke, while drifting closer and closer to the fort. At this point, John could see the trajectory of a cannon ball being launched at his ship. He tried to pull his nearby shipmate out of its path, but he was too late. The mate was disemboweled before his eyes, and another friend was killed right next to him.

During the three-hour battle, three Confederate ships were quickly destroyed or defeated, and the CSS *Tennessee*, a rebel monitor, tried to ram the USS *Brooklyn,* without success. *Tennessee* engaged the entire Union fleet and was overwhelmed and surrendered.

The Battle of Mobile Bay was a victory for the Union but at the total cost of 145 naval deaths and 170 wounded sailors. The *Brooklyn* had 11 sailors killed and 43 wounded, many close friends of John. Twenty-three sailors and Marines on the *Brooklyn* were awarded the Medal of Honor for their gallantry during this battle.

After Mobile Bay, the *Brooklyn* became part of a task force and supported the attack and final surrender of Fort Fisher in North Carolina in mid-January 1865. She sailed north to be decommissioned in late January 1865. At that time, Captain Alden was preparing to leave the ship, and John asked the captain for a favor—one of *Brooklyn*'s old flags. As reported later by North Adams' local paper, *The North Adams Transcript,* John told the captain he "wanted to take it home to the Berkshire hills as a souvenir, and the captain, who said that the Berkshire hills were dear to him, granted the favor" (*Transcript* 5-29-1909).

The Civil War ended in April 1865, and in October, the *Brooklyn* was refitted/recommissioned. For the next three years, she saw service along the Atlantic coast of South America and returned to be recommissioned again in 1870. It is unknown if John remained with the *Brooklyn*, but he was honorably discharged in the early 1870s after nine years of naval service.

In his early forties, John decided to settle down in Adams after traveling the world. He was hired at the Renfrew Manufacturing Company. The company, incorporated in 1867, had at its peak five mills and grew to be Adams' largest employer for almost three decades with 2500 employees. The new mill was built to operate using steam and water power. Its main product was ginghams and yarns, and the company became known for its red-and-white checkered tablecloths. A small village grew up around the Renfrew mills, including a school, row and tenement houses for employees, and a mill store.

A nineteenth-century map of the Renfrew Mfg. Co. complex, including a school and brick row houses on both sides of Columbia Street. *[Courtesy of Adams Historical Society]*

Shortly after beginning work, John met and married Miss Margaret McMillan; presumedly, they chose to live nearby, possibly in Renfrew row housing. With his broad mechanical experience, John was selected to operate the company's first mule-spinning machine at the main mill.

The mule, a large machine used to spin fibers into yarn, was mounted on a lateral carriage. It automated what had been a much slower in-home weaving process, moving at a good rate of speed and needing constant attention. John's position was called a "minder," and several young boys assisted him. Eventually, after a number of years, the process was discontinued, at which point John retired from millwork.

He had enjoyed his years at Renfrew and was popular with the employees. They particularly liked his seagoing stories, as documented by the *Transcript* (*The North Adams Transcript*): "Mr. Welch was one of the most popular of the older employees. During lunch, the younger employees would gather about him and listen to the tales of seafaring life he told. He was full of good stories, and the boys would rather listen to them related from the lips of the old mariner himself than to read them from any book. The boys deemed it to be a great privilege to see the different emblems and pictures that were tattooed on Mr. Welch's body. On his breast is a fine likeness of Admiral Farragut, while on his back between his shoulders is a large American eagle with widespread wings" (*Transcript* 5-29-1909).

When he arrived in Adams, John had joined the newly formed G.A.R. (Grand Army of the Republic) Post 126, named after George Sayles, the first Adams man to die in the Civil War. John would be a faithful lifetime member. The group met at different places in town until the Adams Free Library was built in 1897, then the second floor became designated as their meeting space.

Sadly, Margaret McMillan Welch passed away in 1888 and was buried in the nearby Maple Street Cemetery.

In the early 1900s, John, now in his seventies, retired and moved a dozen or so miles away to Florida Mountain, where he boarded with a Mr. and Mrs. Walsh (no relation) and kept a robust schedule. A news article in the local paper notes, "Daily Mr. Welch walks a mile and a half a day up the mountain and then returns to his home. He walks uprightly and says the mountain air is the greatest of all tonics. He is also a great swimmer. He learned to swim in the Hoosic River at Williamstown as a boy, and still keeps it up, going frequently at Haskins Pond in Florida. He says he can swim more easily than he can walk" (*Transcript* 5-29-1909).

The seasoned mariner, reflecting on his colorful past. *[Courtesy of Adams Historical Society]*

John continued to attend G.A.R. meetings in Adams, and in 1909, just before Memorial Day, he presented the Post with the USS *Brooklyn*'s massive battle flag that he so long treasured. The 24 foot by 14 foot honored relic was initially displayed behind the commander's chair in their second-floor meeting room. Not long after John died in 1915, the flag was placed in a glass case, and since 2005 the historical flag has been stored in appropriate archival grade materials. It was displayed only once, in 2014.

John remained active well into his eighties, had successful cataract surgery on his right eye at eighty-two, recovered from a critical bout with pneumonia at eighty-three, and continued to swim and bathe in Haskins Pond until his death on December 30 or 31, 1915. At that time, the faithful mariner was buried with his wife Margaret at Maple Street Cemetery. In John's will, he left his "entire estate [of] $1,139.03 to the G.A.R. to be 'used for charitable purposes.'"

The USS *Brooklyn* flag on display in Memorial Hall of the Adams Free Library, 2014. *[Courtesy of Adams Historical Society]*

The gravesite of John and Margaret Welch, in the shadow of the historic Quaker Meetinghouse, built in 1784. *[Courtesy of the author]*

A journey that began when a stripling lad learned to swim in the Hoosic River in Williamstown became something far greater: The lad sailed across the world in merchant vessels, survived countless storms and a shipwreck, then participated in two of the Civil

War's most consequential Naval battles. His journey may have ended peacefully in the small town of Adams, where he entrusted his memories and treasures to his fellow veterans, but his adventures survive in story—a tradition of all great seafarers.

Major Reuben A. Whipple

CIVIL WAR, SPANISH-AMERICAN WAR, AND PHILIPPINE INSURRECTION

He was from a long lineage of soldiers and sailors who had served their country, but few in the Berkshires realized that "Major Whipple," as he was known to all, had been one of few men to serve heroically in three of America's wars, from enlistment in the Union Army in 1864 to his return home from Manila in 1901.

All three wars brought great peril, with opportunity for courageous feats, but one clash stood out in particular. In the battle for the city of Santiago de Cuba, during the Spanish-American War, Whipple survived what was described as a "hailstorm" of bullets as he urged his men onward, and they seized their objective. Above all, Major Whipple was known for always leading from the front.

REUBEN ARNOLD WHIPPLE WAS born in Smithfield, Rhode Island, on August 11, 1846, to George and Sarah (Wescott) Whipple. The family moved to the Zylonite section of Adams in 1853 when Reuben was seven years old, and he attended local public schools. He finished his studies at The Institute in Lanesboro.

His dad bought and co-owned Follett & Whipple, the predecessor of the New England Lime Company. George owned lime-burning kilns in Adams, North Adams, and in Pownal, Vermont. Reuben worked there part-time and, in the summers, learned about his dad's business, where he would work later in life.

As a youth, he much preferred the outdoors, wandering the

nearby hills and spending his time hunting and fishing. He had an affinity for horses, and his six-foot-two-inch frame could often be found straddling a friend's horse at a nearby farm.

After completing his elementary education, Reuben worked full-time at Follett & Whipple until enlisting in the Union Army in 1864 at age 18. Private Whipple joined Company B of the 8th Massachusetts Infantry, which belonged to the 3rd Separate Brigade of the 8th Corps.

He protected the Northern Central Railroad from predatory attacks by Confederate guerillas. After slight skirmishes and other assignments, his company was assigned garrison duty in Baltimore. Eventually, Private Whipple's Company B returned to Massachusetts and was mustered out of service in November 1864. The unit had no combat casualties but lost 15 men to disease. As records indicate, "He served faithfully until the close of the rebellion . . . ."

Notably, Reuben was a descendant of Commander Abraham Whipple, who, during the French War, commanded a celebrated privateer called the *Game Cock*, which captured 23 French ships during a single cruise. Promoted to Commodore, Abraham also destroyed the first British vessel in the Revolutionary War, the cutter *Gaspee*, and led a squadron of US ships before the end of the war.

War, and success in it, seemed to follow the Whipples.

Returning home, Reuben resumed working with his dad, then married Evelyn "Eva" Todd in 1871. He had met Eva on Park Street in downtown Adams one afternoon. Her father owned a nearby shoe factory. The happy couple would go on to have four children, two boys and two girls—one whose life would end tragically.

Reuben, always interested in civic affairs, served three years as a selectman during the time the town was divided into Adams and North Adams in 1878.

In 1887 when Company M was created as a voluntary militia of Massachusetts, Reuben was elected the first captain of the newly created forty-four-man company. The group quickly formed a rifle team that became celebrated throughout the state and was highly competitive.

Captain Whipple created high standards for Company M's rifle team. He was an excellent marksman and, for years, led the team. He and his team won many buttons for outstanding marksmanship. For eight years, he remained in Company M, but in 1894 he was appointed major of the 2nd Battalion. Company M lost one of the best-liked captains in the volunteer militia. In his new position, the major served faithfully until 1898, when he was commissioned a major of the volunteers in the same regiment.

Portrait of Major Reuben A. Whipple, from the officers' roster, circa 1894. *[As reproduced in* Company M and Adams in the War with Spain, *Herbert O. Hicks and Fred A. Simmons, 1899]*

Company M continued to grow in his absence and built a new armory in Adams in 1895.

An interesting article made the newspaper on July 11, 1896, titled "A Narrow Escape." It begins, "Major R.A. Whipple Saves the Lives of Mrs. E.J. LaFerriere and Daughter." It was on the switch track near his father's lime kiln business when the horse pulling Mrs. LaFerriere's buggy panicked and plunged directly into a train's path. The major "had to exert all his strength to hold [the horse] as the train went by at a high rate of speed . . . . The front wheels of the carriage were broken into pieces . . .[but] they escaped without injury."

The Approach of the Spanish-American War

By the late 1800s, Spain had ruled Cuba for many years, in some cases violently suppressing Cuba's yearning for independence, according to American papers. Americans saw similarities to their own struggle with Great Britain during the Revolutionary War and developed a sympathy for the Cubans.

When violence continued in Havana, with the Spanish Army trying to quell the insurgency, the American government sent the USS *Maine*, a battleship named for the state of Maine, to Havana harbor in hopes of protecting American citizens living in Cuba.

Just two weeks after its arrival, on February 15, 1898, while docked at harbor, the *Maine* was destroyed by an explosion, and 268 crew members were killed while they were sleeping. Five tons of explosives intended for its 6- and 10-inch guns detonated. Disagreements continue to this day about the cause of the explosion, ranging from an enemy mine/torpedo to internal gases.

Americans were outraged, and a little over two months later, on April 20, 1898, stirred by national newspapers, military force was authorized to help the Cuban patriots, and war was declared. Americans headed to Cuba to help Cubans fight for their independence, although many thought the insurgency was already close to establishing independence without outside help.

(The USS *Maine* lay at the bottom of the harbor until 1911 when it was raised, towed to sea, and sunk. It now lies 3600 feet below the surface. The ship's main mast was conserved to stand as a memorial at Arlington National Cemetery.)

Before declaring war, the United States offered to buy Cuba from Spain; neither Spain nor the rebels wanted to give up the land. Ultimately, President McKinley called to mobilize 125,000 volunteers. Remember the *Maine*, To Hell with Spain became a famous slogan. Americans would fight for four months and suffer under four hundred casualties, though thousands of additional men would be lost to disease.

Adams residents, like the rest of the country, were stirred up, and Major Whipple and Company M prepared to leave for Cuba. This time, his son, Corporal George E. Whipple, would be accompanying the major.

Company M's officers were given a physical exam, and according to a news article, some thought Major Whipple would be rejected because he had false teeth. When he appeared before the medical board, the major had his teeth in his pocket, and the absence of teeth was remarked upon by the surgeon. The major replied that he was in the business of fighting Spaniards, not eating them. He was passed (*Pittsfield Eagle* 5-13-1898).

Company M's eighty soldiers, part of the 2nd Massachusetts Regiment, left Camp Dewey in Framingham in May 1898, outfitted at the arsenal with their Springfield Model 1892 Krag-Jorgensen rifles. They were headed for Tampa. Company M, 2nd Massachusetts Infantry, was the first volunteer company in the United States to leave its armory for the Spanish War.

They trained via Jersey City and then Washington, DC, occupying three train cars. Their subsistence was very basic; hard crackers, corned beef, and coffee three times a day. One of the volunteers said he "understands why the surgeons are concerned about their teeth."

Upon reaching Tampa, the company regrouped. During this time, Captain Whipple purchased what news articles called an "Arizona pony" at a local stockyard, and it would accompany him to Cuba. The horse was a handful. Being described as a pony does not seem apt. It was sixteen hands tall, a fiery, temperamental, black thoroughbred. Captain and horse (what will soon become known as his "Santiago horse") seemed to reach an understanding.

Company M was loaded on the "Manteo" transport, each man with thirteen days' rations, and began sailing to Cuba on June 13. In late June 1898, Company M waded ashore on the beaches of Daiquiri to begin the trek westward toward the primary goal of capturing the city of Santiago de Cuba, fourteen miles away. Daiquiri would continue to serve as a resupply depot during their time in Cuba.

They skirmished with Spanish soldiers all along the way, continually pushing them back toward the primary objective: capturing Santiago de Cuba. Company M supported the capture of the town of Seville as they went, and as they closed in on Santiago, the fighting stiffened.

On July 1, 1898, Company M took part in the devastating attack on the fortification at El Caney, where 500 better-armed Spanish soldiers held off 6000 American troops. It's there that Company M had its first man die from wounds received in action: eighteen-year-old Private Alfred Thiel from Adams. Young Thiel was found with a severe scalp wound and sent to the rear, with hopes of saving him. He returned home, although he continued to decline and soon died from a complicating fever (*Transcript* 8-4-1898).

Company M was worn out by the previous day's long march at El Caney, plus the soul-sapping heat. They faced an uphill long line of entrenched Spaniards using what was described as "smokeless Mauser rifles," sending repeated, well-aimed volleys of shots into their regiment.

Also known as the Spanish Mauser model 1893, this rugged, dependable bolt action rifle featured a magazine that could be quickly reloaded with five smokeless rounds of 7x57mm ammunition. It was considered one of the most successful designs ever produced. The Mauser was considered a superior weapon to Company M's Springfield, specifically in its ease of use and rate of fire. Soldiers often found themselves at a disadvantage and unable to match the speed of fire given off by the Mausers.

Some of the American forces, including part of the 2nd Massachusetts Regiment, to which Company M belonged, did not have the superior Krag-Jorgensen rifles and were using old trap-door, black powder, Springfield rifles, creating a lot of smoke and making them easy targets for the Spaniards. (The M1903 Springfield eventually replaced the obsolete rifles in 1907.)

When two volunteer regiments were ordered to the rear because their black powder rifles were attracting enemy fire, the New York regiment promptly left. Company M stood fast, and those having black powdered rifles simply looked for opportunities to fetch modern smokeless rifles from the men fallen all around them.

The Americans ended up capturing the fortification and, moving on, focused on the city of Santiago de Cuba. However, El Caney had cost them 1300 dead and wounded, while the Spaniards had fewer than 200 casualties.

The Santiago battle and siege started just days later; Major Whipple was at the center of the action. During the two-week siege, *The North Adams Transcript* (*Transcript*) quoted a soldier as saying, "'During the battle [of Santiago] Major Whipple [and two other officers] stuck their swords in the ground and stayed right in the center of Company M while bullets fell like hail stones . . . its officers were as brave soldiers and as good men as ever were in battle.' They held their position through the ten-hour fight at Santiago, winning the admiration of the seasoned veterans. He also remarked that Major Whipple and Lieutenant LaFerriere had grown old fast. He said their hair was almost white" (*Transcript* 7-20-1898).

Another trooper was quoted as seeing "Major Whipple right at the front, yelling Give 'em Hell, Boys" (*Transcript* 7-27-1898).

Santiago was captured on July 17, 1898, and although some resistance continued, the capture of this significant city essentially ended the war. Spain sued for peace in July and signed a peace protocol in August, relinquishing all rights to Cuba.

Company M saw two months of almost continuous action and occupation duty. The Cuban campaign had been brutal. The terrain was difficult to negotiate, and the weather was humid and draining. Supplies were difficult to obtain. Food was scarce; at one point, some units ate their horses' oats. Artillery support was lacking. Some old-time commanders still used frontal assaults that significantly increased casualty rates.

Mule trains seemed to run continuously from Daiquiri with ammunition and supplies, but there needed to be more, and they always seemed to arrive after the battle when an earlier arrival would have been most beneficial. The trains were also pressed into use to bring the numerous wounded back from the battlefield, though they were often subject to snipers, slowing return visits.

There was a significant amount of criticism of regimental officers, but none for the officers on the ground with the soldiers. Major Whipple is said to have lost 40 pounds during the two-month campaign.

Corporal George Whipple, the Major's son, was hospitalized with malaria during the end of the campaign. His condition worsened, and he died suddenly on August 12, 1898, just days before Company M embarked for the United States. Major Whipple was devastated. The twenty-five-year-old soldier, an 1891 Adams High graduate, an all-around athlete and a boxer, was popular with the troops. He joined Company M when he was 15 years old and was one of the company's best marksmen. Corporal Whipple had survived all the fighting, only to die from disease. He was buried along with the regiment's officers in a tribute to his dad.

Company M returned home in late August 1898 to a welcoming crowd at the train station. A few days later, Major Whipple "received the Arizona pony which he had at Santiago. . . . [The pony] is not suffering from malarial fever and seems in pretty fair condition" (*Transcript*, 8-30-1898).

As part of the local celebration, a local druggist, J. Wells Thompson, is noted in a newspaper article as having "Attractive Show Windows . . . .contain[ing] a number of war relics. . . . A Krag-Jorgensen carried by Corporal George E. Whipple, a Mauser owned by Major Whipple, and a calvary rifle. The cartridge belts worn by Major Whipple and his son, Corporal Whipple. In the latter's belt is part of a shell that exploded, having been hit by a bullet . . . and a Spanish sword presented to the Major by a Spanish officer."

As a result of Major Whipple's battlefield actions, he and several other field commanders were recommended to be "brevetted," an honor accorded for gallantry in the field. This honorary title was granted. (Today's military has replaced this honor with the bestowing of medals for gallantry in action.)

While George's death and the recovery of his body remained on the Major's mind, he and his Santiago horse were in demand by residents wanting to celebrate winning the war with Spain. He and the pony were often the centerpiece at local parades and were the major attraction at the Pittsfield 1898 Labor Day Parade.

Major Whipple leading the troops in the Labor Day Parade. *[Courtesy of Adams Historical Society]*

Finally, in January 1899, the Major secured permission to return to Cuba with a delegation to retrieve the bodies of America's war dead. The delegation had the task of locating over 600 bodies and bringing them back home. The major looked for ten bodies from his regiment, including George's.

Of the 686 soldiers found, 100 could not be identified. All were located in rude trenches between San Juan and El Caney, where most of the fighting took place. Sometimes when graves had been opened, there would be five or six bodies, rather than finding two as the headstone would indicate.

The 2nd Regiment's dead were better marked. Inside the pocket of each soldier was placed a bottle containing a slip on which was written his name, company, and regiment, and not a man was buried without one. The younger Whipple's read, "Corporal George Whipple, son of Major Whipple, was buried with the officers, his father being a major" (2-27-1899).

After several months of work, the transport ship carrying the bodies headed to America. The trip was described by a passenger, Doctor Field, in an emotive if not entirely accurate news article as follows:

***Our Ship of the Dead—Trip of the Transport* Crook**

> *The United States transport* Crook, *with 686 bodies of the nation's dead heroes from Cuba and Puerto Rico, anchored off Liberty Island recently. . . . [Dr. Field] said, "There was one father who was escorting the body of his soldier son from the battlefield of San Juan to the boy's home up in the Connecticut Valley. I think he was Major Whipple of the 5th Brigade, himself belonging to the 2nd Massachusetts Volunteers and coming from Deerfield. The Major's son, Lieutenant George E. Whipple, Company M, Second Massachusetts, had passed through the campaign unscratched but after four days of illness at Santiago on the day before his Regiment was to sail for old Massachusetts, the boy died. This was on August 12 . . . . I shall never forget the melancholy satisfaction with which Major Whipple was returning with his sacred dust."*

On April 1, four bodies arrived in Adams to lie in state at the Massachusetts Armory under an honor guard. The coffins were covered with flags and wreaths while many townspeople paid tribute. The Saturday procession, accompanied by Company M, started from the armory and proceeded to Maple Street Cemetery, where Corporal Whipple was buried in the family plot. At each grave, a salute was fired, "Taps" was played, and a chaplain conducted services. Town flags flew at half-staff.

It was the saddest event the town had seen, since those wounded during the war returned home, died from their wounds, and were so much honored. George's name is perpetuated in the Rusek-Whipple Memorial Square at Center and Crandall Streets.

There would be little rest from battle for Company M. At the end of the Spanish-American War, Spain ceded the Philippines to the United States, and Filipinos who sought their independence rebelled against having new occupiers. The first battle, which Americans called the Philippine Insurrection, occurred just months after the peace treaty was signed for the Spanish-American War.

To our modern-day sensibilities, helping one island nation secure independence just to receive another island nation as spoils of war seems hypocritical, or at least perplexing. In some cases, paternalistic Americans thought the Philippines was incapable of ruling itself; in other cases, hawkish Americans worried that other countries might occupy it if a power vacuum was allowed. The resulting two-and-a-half-year war was merciless. Hundreds of thousands of civilians, 20,000 Filipino troops, and over 4000 American soldiers would perish from violence, famine, and disease.

In June 1899, the fighting erupted between the United States and the Philippines, and war was declared. The same month, President McKinley granted a commission of Captaincy in the regular Army to Major Whipple, a reservist.

Now Captain Whipple was charged with recruiting and forming the 26th Regiment for immediate service in the Philippines. Recruiting efforts begin in Springfield, Massachusetts. Many applicants tried to ensure they would be placed in Captain Whipple's company, which as an auxiliary Army unit was heading to Manila as soon as it was fully formed. His company was to have 125 men.

There was high interest, and requirements were lowered: "Usual requirements of citizenship and the ability to read and write are waived, and men just able to speak English are accepted. All need to be between 18 and 35 years old" (*Transcript* 7-10-1899). Captain Whipple oversaw the efforts, helped by a physician.

It took some months to fill all the 26th Regiment's positions, then the unit underwent three months of training in Plattsburgh, New York.

The regiment was mustered in on September 6, 1899, in Plattsburgh and started for Manila by way of Boston. Then it was on to Providence and San Francisco, finally embarking on the transport *Grant*, reaching Manila on October 24.

The regiment, without disembarking, was ordered to Iloilo on the island of Panay. The United States conducted nine military campaigns; one was the Iloilo campaign. Captain Reuben Whipple and Company M would arrive shortly after the city of Iloilo had been seized in a vicious battle with Filipino forces and left in a blackened ruin, destroyed by arson, naval bombardment, and street fighting.

For the next eighteen months, the 26th Regiment took part in hard campaigning on Panay against the rebels. Captain Whipple found himself leading his company in a perilous environment filled with daily ambushes, and while the rebels were ill-equipped, they were canny, devoted to their cause, and knew the terrain.

Captain Whipple, who served as Company M's quartermaster, resigned when he accepted another captaincy in the regiment. The Captain temporarily assumed command of Company L, an infantry company, so that he could get into the fight. His son, Private Robert Whipple, joined him in Company L.

It was no-quarter warfare. There were sharp, intense, brief engagements for soldiers of Company L and M: close-in, face-to-face fights with Bolo knives, bows and arrows, and single-shot antique guns against bayonets and American Springfields.

Once there, a fight ensued against dug-in, strongly fortified Filipino forces under the command of Captain El Berto in the mountainous regions of Iliolo. A news article from the *Transcript* (3-5-1901) reports that 50 soldiers from Company M garrisoned at a small town named Igbaras, led by native guides, found themselves under concentrated fire from three sides as they were set to burn down a defensive barrio to smoke out the rebels.

The soldiers found the insurgents were well placed, entrenched in almost impregnable positions. The men retreated, carrying their wounded men, both from North Adams.

The Philippine Insurrection was noted for many atrocities committed by both warring parties. Wounded prisoners were often killed. Villages were relocated, and inhabitants often starved to death. Extrajudicial executions occurred in retaliation for soldiers' deaths.

A number of investigations concluded that American soldiers "water-cured" rebel leaders and burnt villages down, many at the direction of their commanders.

After serving eighteen months in combat, the 26th Volunteer Regiment and Company M returned to the United States. General MacArthur ". . . notified the war department that the transport *Garonne* sailed from Manila today for San Francisco with 41 officers and 849 enlisted men . . . ." (*Transcript* 3-11-1901).

The trip home took over a month, after which the regiment was mustered out.

Major Whipple and his son Robert returned home safely in late May 1901. Before their arrival, "A large box was recently received from the Captain, containing knives, bolos, daggers, and various other native weapons such as our soldiers have had to face. These are interesting curiosities but no match for American rifles" (*Transcript* 5-21-1901).

With the undeniable military and equipment advantages and the capture or surrender of key rebel leaders, the Philippine Insurrection ended thirteen months later in July 1902, with President Roosevelt declaring a general amnesty and establishing provincial governments.

"Major," as he was known to everyone, served the Town of Adams as a Deputy Sheriff for several years. He was reappointed by the town selectmen and held the positions of Superintendent of Streets, Parks, and Buildings and was Superintendent of the Fire District until his passing.

As the years passed, many of his friends solicited Congress to pass a bill awarding the three-war veteran a pension. They requested that he be retired "on half pay [as] the regular army officers who served in the Civil War, the War with Spain, and the War in the Philippines" (*Transcript* 3-10-1902). Only two officers in the State of Massachusetts qualified for this pension. Records are unclear as to whether or not the bill was passed.

At the age of 61 on October 13, 1907, after several weeks of illness, at the retired rank of Lieutenant Colonel, Reuben Arnold Whipple, better known to all as Major, died at his home on 44 Summer Street. One of the few survivors of three wars ultimately succumbed to liver disease.

Military representatives from across the state, town officials, family, and friends gathered at St. Paul's Universal Church in Adams to honor the memory of their comrade, this most unique man who had given his life to his country and community, always committed to what more he could do to help.

CORPORAL HENRY F. CARON & SERGEANT HENRY L. CARON

UNCLE/NEPHEW

WORLD WAR I & WORLD WAR II

Henry F. Caron and his nephew Henry L. Caron were from the same family, a generation apart. Both joined the same small-town National Guard unit, serving in two different wars. Both became decorated heroes. One made the ultimate sacrifice, while facing down hundreds of advancing enemy soldiers in an overwhelming force. The other sacrificed in perhaps a more traumatic way, suffering years of prolonged, torturous captivity thousands of miles from home. Their stories deserve to be told as one, influencing and inspiring each other.

IN 1896, HENRY F. CARON was born in Adams to John Peter and Amanda (Hudon) Caron, immigrants from Canada in the late 1800s. Henry was born at home, as were his sister and brother.

His dad worked as a local Teamster, and his mother worked in the nearby mills until her unexpected death when Henry was only four years old. With little support, Henry's dad sent him to live with his aunt and uncle, where he was educated at Notre Dame parochial and Liberty Street schools.

After he left school, Henry learned his craft as a carpenter, working for the Boston & Albany railroad. While employed with the railroad, he enlisted as a teenager in the Adams National Guard–Company M and served three years from 1912 to 1915.

Company M had been in existence since 1887. Before that time, various militias had responded to local needs. The State Armory, a bold fortress-like presence, sat close to the town square and was constructed and dedicated on April 17, 1914. (It is no longer in use.)

Life in the small town of Adams was shaped by the closely woven influences of family, school, church, and Company M. They mostly were located within a stone's throw of each other. The Adams community was deeply proud of Company M. It had a solid reputation as a State Guard unit and a championship rifle/machine-gun team. Company M would invite the town's residents to "Inspection" and "Military Nights" where they could view and handle military hardware and watch the soldiers drill. Company M sponsored military balls that were well-attended, gala affairs. The community also loved to see its troop march in the Memorial Day parade.

When trouble broke out at the Mexican border, Henry reenlisted, as *The North Adams Transcript* noted. Company M was called into service on June 19, 1916, for duty in Columbus, New Mexico, to support General Pershing's pursuit of Francisco "Pancho" Villa, a marauding cross-border guerilla who had attacked camps and killed a number of Americans.

On June 21, 1916, Company M deployed to New Mexico after President Wilson called up 140,000 National Guard troops. When Captain Potter, Company M's commander, and 108 soldiers, including Corporal Caron, left Adams by train on June 21, "there were more than a thousand people at the railroad station to give the boys a sendoff" according to the *Transcript.* After further training and travel, Company M arrived in New Mexico on July 1 and guarded the punitive expedition's main base until returning home in late October 1916. Villa was never captured.

Just four months later, on February 20, 1917, Henry and Company M were called out again to guard local government and railroad infrastructure after relations with the German government worsened, and diplomatic ties with Germany were severed. The men were roused at midnight by phone or messenger and spent the night at the armory; fellow soldiers living in North Adams arrived by trolley the next day, and then the group was deployed across western Massachusetts.

Corporal Caron was one of the noncommissioned officers in charge of the thirty troopers called out to guard a bridge over the Deerfield River as well as the east and west portals of the Hoosac Tunnel. At the time, rumors spread throughout Adams that war had been declared, though in fact that did not occur for six more weeks, on April 6, 1917.

This portrait of Cpl. Henry F. Caron is the only photograph known to exist of the World War I soldier from Adams who lost his life in the war. *[Courtesy of Adams Historical Society]*

Shortly after guard duty ended, Company M began training for an overseas assignment. It was transferred to federal service and became Company M of the 104th Infantry Regiment, Yankee Division. After more training and much travel, Corporal Caron's 250-man company landed in France in early November 1917. After three more months of training by French troops, they were entrained and marched miles to the front lines in late February 1918. In early spring, the company had its first non-combat death: Phillip Callery was accidentally killed when he fell against his bayonet (*Transcript* 4-15-1918).

For the next nine months until the Armistice in November, Company M, except for a few brief respites, would be fighting for its life in some of World War I's most bitter, bloody, and contested battles. When Company M saw combat, it faced devastating weapons unseen in previous wars. The men encountered huge, heavily armored tanks, rapid-fire machine guns, heavy artillery, invisible yet deadly poisonous gases—and, overhead, airplanes equipped with bombs and machine guns.

Company M initially occupied front-line trenches in an area north of Soissons called Chemin-des-Dames, then moved to the Toul front near Bois Brule (burnt woods), and was finally placed in trenches near the small village of Apremont. Pictures of the landscape show barren moonscape ridges, too many artillery shell holes to count, shattered spindly trunks of trees, totally denuded terrain, and exhausted, thin, glassy-eyed soldiers, many wearing googly-eyed gas masks.

Apremont was central to the whole defensive system of the area and highly sought after by the German high command. Company M's trenches and the German defensive works were, in some places, just 40 yards apart. Because of the proximity and the hard-fought battles for the area, the entire section became known as the "suicide sector."

On the front line for over forty days, but in this section only recently, Company M, part of the 3rd Battalion, was barely prepared for the ensuing savage fight over this scarred piece of land.

Leaders of the 25th, 36th, and 65th German regiments had decided they needed a decisive breakthrough, and Company M occupied the desired land. The Prussian commanders handpicked their most brutal 800 soldiers for the job, shock troops (special raiders) known as *Hindenburg's Traveling Circus.* They were the strongest, best equipped, most experienced, and gave no quarter.

On the night of April 9, 1918, the Germans unleashed a barrage of poison gas and high explosives that covered Company M's trench areas. Each side could hear the other troops gathering in the thick predawn fog. Anxious stretcher bearers waited nearby, preparing to collect casualties, and everyone was anticipating the dreaded whistle that would start the melee.

It came suddenly, almost unexpectedly, and the attack began; the prevailing winds had cleared the poisonous gas except in the deepest part of the muddy trenches. The men advanced, mud clinging to their boots, wearing heavily soddened uniforms, with bayonets at the ready and truncheons on their belts.

The battles were individual and vicious. Attacking troops clambered over barbed wire covered by interlocking machine guns and quickly jumped into trenches being defended by soldiers standing atop earthen parapets that allowed them to fire over the top sandbags. The attackers slid into muddy quagmires, embracing other muddy figures fighting for their lives in bloody, hand-to-hand battles that were usually finished with a knife.

After repeated efforts, Hindenburg's shock troops initially secured a foothold in part of the trenches, but after strong counterattacks by the 104th Infantry, including Henry's Company M, they were repelled.

It was during these counterattacks with his platoon, reported the *Transcript*, "that [Henry F. Caron] was attacked from behind while hurling grenades at the 'Boche' who were in the trench ahead of him, and it is evident the trench he was in must just been taken by the Allied forces and the enemy was still in the trench" (4-30-1918).

It is unknown whether he was bayonetted or shot in the back, although it appears his comrades quickly dispatched his assailants and nurtured him during his last moments.

The same news article continues: "The report which comes from France states that 'with courage and remarkable devotedness on April 10 when mortally wounded he passed the remainder of his grenades to his comrades saying, *I cannot use these, go to it.*'"

The line held. When the battle ended, Company M and the 3rd Battalion prevailed, with 50 dead and wounded soldiers compared to the Germans' 600 lost men. Corporal Henry L. Caron was the first member of Company M to meet his death in combat. In a sad coincidence, his death occurred on the date of the first anniversary of his marriage to Mary Rosch from Adams.

Apremont was one of the first major battles of World War I involving American troops, and they had acquitted themselves quite well. Several weeks after the battle, a French general gathered the 104th Regiment and members of Company M, removed from his own uniform his personal decoration, the Croix de Guerre, and pinned it on the regimental colors. He stated how proud he was to pin it on a regiment that had shown such fortitude and courage. It was the first time that an American military unit had been decorated by a foreign government in World War I.

Henry himself was awarded the Croix de Guerre posthumously by the Commanding General of the French Army corps with whom his regiment was serving at the time. The general, when forwarding the award to Henry's wife, wrote that he took "pleasure in sending you [Mrs. Caron] the cross herein to retain as a permanent mark of the recognition shown of his courage and devotion to duty" (*Transcript* 7-2-1918).

Mrs. Caron did not receive Henry's medal immediately. The medal had been forwarded to his uncle, still listed as Henry's next of kin from service on the Mexican border. The legal designation of beneficiary had changed upon Henry's marriage, but paperwork was misplaced. It took eight months for the medal to be surrendered by the uncle and given to Mary.

The 104th was later considered a second time to receive the Croix de Guerre, again for undaunted bravery, but the decoration can only be awarded to a unit once. It is interesting to note that during World War I, only 117 soldiers from the entire 104th Regiment were decorated with the Croix de Guerre, with Corporal Henry F. Caron the only soldier from Adams receiving this highest honor.

The honor of receiving the Croix de Guerre was celebrated by the 104th Regiment every year thereafter, and one of those reunions was held in Adams in 1937. At that reunion, the town decided to change the name of Kipper Avenue in the Zylonite section to Apremont

Street, in honor of the many local men who had fought there.

After Apremont, and the loss of Corporal Caron, Company M continued the fight and participated in major battles at Belleau Woods, Chateau Thierry, assaults near Verdun, and were still engaging the Germans when Armistice was declared on November 11, 1918.

After months of distinguished service, the company returned home by boat and was released from federal service at Camp Devens on April 26, 1919, seventeen months after landing in France. The casualty record of Company M was listed in *The North Adams Transcript* on April 5, 1919:

124 Casualties: 20 Killed in Action, 73 severely/slightly wounded, 4 shell shocked, 21 Gassed, 1 died non-combat, 1 died of disease, and 4 died of self-inflicted wounds.

Corporal Caron, initially interred in France, was returned home in April 1922, and a wake was held at his family home in North Adams. American Legion members escorted his body to St. Francis Church; Henry's squad members from his time in France acted as bearers. Burial was conducted with full military honors at St. Joseph's Cemetery.

The Town honors the memory of Henry Caron on Memorial Day with flags and wreaths. *[Courtesy of the author]*

Another honor accorded Corporal Henry Caron was when the Town of Adams dedicated Henry Caron Square at the intersection of Center, Commercial, and Myrtle Streets where they converge near the Post Office.

The United States fought in World War I for nineteen months. Company M, the pride of the Berkshires, spent nine of those months in direct combat and lost a brave fighter in Henry Caron, Adams' first combat death and its first World War I hero.

HENRY LAWRENCE CARON WAS born to Albert and Albina Caron in 1920, named after his uncle, the recent war hero. The newborn arrived almost two years after his famous uncle's death.

Albert Caron worked many years at LL Brown Paper Company and retired from General Electric in Pittsfield, Massachusetts. His wife Albina was a diligent homemaker who also worked at Brightwater Paper and Sprague Electric Companies.

The Caron family lived on Glenn Street as it headed north out of Adams toward Cheshire. They had four children, three boys and a girl, spaced over twenty years. Henry could always be seen tagging along with his older brother Albert and his friends. The boys spent a lot of time in the woods and across the street playing in the Hoosic River.

When Henry was nine years old, he and a group of friends made a game out of crossing and recrossing the busy nearby Commercial Street, only for Henry to be hit by a car. It was exciting enough to make the papers: "Taken into the Little gas station . . . . Dr. McLaughlin was called. It was found that no bones were broken and that he had only been shaken up and bruised" (*Transcript* 3-28-1929).

Henry, a tall, lanky, broad-shouldered boy, attended nearby Commercial Street Elementary School where he applied himself and was recognized for penmanship by the Palmer Company. He subsequently attended Liberty Street School.

As a teenager, he worked at odd jobs around town, usually turning over his earnings to his mom, keeping a small portion for himself.

Henry's main thought growing up was joining Company M as soon as he turned seventeen. He wanted to follow in the footsteps of his well-known uncle. In 1937, he enlisted, was welcomed at the local National Guard Company, and began training immediately.

After eighteen months of local service and a desire to see more of the world, Henry requested to be honorably discharged so he could join the Marine Corps. He entered the Marines in August 1938.

After Parris Island boot camp, Private Henry Caron was stationed at Dover, New Jersey, and then sent to the Marine detachment in Shanghai, China, for the next eighteen months. In autumn 1941, just two months before the attack on Pearl Harbor, he was transferred to Cavite Naval Base in the Philippines.

Pearl Harbor was attacked on December 7, 1941, and the Philippines was attacked the next day. When the surprise attack occurred, Henry was stationed at Fort Hughes, which is located on Caballo Island just off the coast from Cavite Naval Base.

The fort was isolated in the center of Manila Bay off of Corregidor Island and was intended to be part of the Harbor Defenses. After five long months resisting the Japanese, the fort and its defensive artillery batteries surrendered on May 6, 1942. Both Fort Hughes and Corregidor surrendered to the Japanese at the same time.

Weeks later Henry's parents received the following Postal Telegraph:

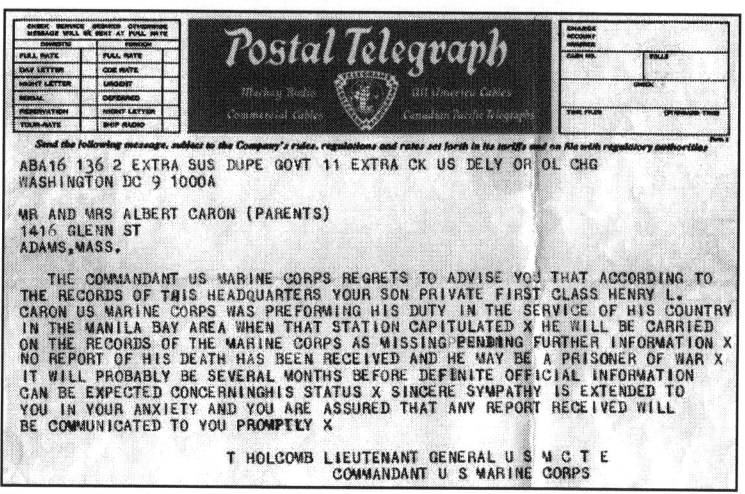

Postal Telegraph

ABA16 136 2 EXTRA SUS DUPE GOVT 11 EXTRA CK US DELY OR OL CHG
WASHINGTON DC 9 1000A

MR AND MRS ALBERT CARON (PARENTS)
1416 GLENN ST
ADAMS,MASS.

THE COMMANDANT US MARINE CORPS REGRETS TO ADVISE YOU THAT ACCORDING TO THE RECORDS OF THIS HEADQUARTERS YOUR SON PRIVATE FIRST CLASS HENRY L. CARON US MARINE CORPS WAS PREFORMING HIS DUTY IN THE SERVICE OF HIS COUNTRY IN THE MANILA BAY AREA WHEN THAT STATION CAPITULATED X HE WILL BE CARRIED ON THE RECORDS OF THE MARINE CORPS AS MISSING PENDING FURTHER INFORMATION X NO REPORT OF HIS DEATH HAS BEEN RECEIVED AND HE MAY BE A PRISONER OF WAR X IT WILL PROBABLY BE SEVERAL MONTHS BEFORE DEFINITE OFFICIAL INFORMATION CAN BE EXPECTED CONCERNINGHIS STATUS X SINCERE SYMPATHY IS EXTENDED TO YOU IN YOUR ANXIETY AND YOU ARE ASSURED THAT ANY REPORT RECEIVED WILL BE COMMUNICATED TO YOU PROMPTLY X

T HOLCOMB LIEUTENANT GENERAL U S M C T E
COMMANDANT U S MARINE CORPS

[Courtesy of Murphy/Caron family.]

The prisoners from Fort Hughes and nearby Cavite Naval Base were captured in May 1942, a month after the infamous Bataan Death March, although the same type of brutal treatment prevailed for these soldiers and sailors. The men were beaten, stripped of their possessions, and anyone holding Japanese money or war souvenirs was often killed, presumed to have stripped or slain Japanese soldiers.

Herded together, transported to the mainland, the group was forced marched at the point of a bayonet, with no food and little water (except roadside puddles) the 20 miles to Manila. Upon arrival, they were packed into a converted civilian prison called Bilibid. Henry mentions it in a retrospective *Transcript* article dated October 27, 1945, recalling marching for two or three days before being incarcerated. (Bataan Death March captives were forced to cover 60 miles in the unrelenting sun.)

Bilibid was used as a transit center, and almost all POWs passed through there at one time or another. After several months, for some reason Henry was transferred north to Cabanatuan prison and remained there until his return to Bilibid before being shipped to China.

Conditions in both camps were abysmal. There was little shelter, hardly any food, and no medical treatment. Disease was rampant, deaths frequent, and only the fortunate, hardy, and very lucky survived.

After spending six months in the Japanese POW camps of Bilibid and Cabanatuan, Henry's 170-pound, lanky frame was now down to 120 pounds. When he returned to Bilibid in Manila, he was brought to the seaport and packed with hundreds of others on what was known a "hell ship" and headed for China.

In 1942, Japanese began transferring POWs by sea in the packed holds of cargo ships to other countries to be used as slave laborers. Crammed into holds with little food, water, and ventilation, many died during the weeks it took to get to their destinations. Also, many of the unmarked "hell ships" were transporting enemy soldiers and thus sunk by Allied planes. It is estimated that over 20,000 Allied POWs were lost at sea on ships bombed by friendly forces.

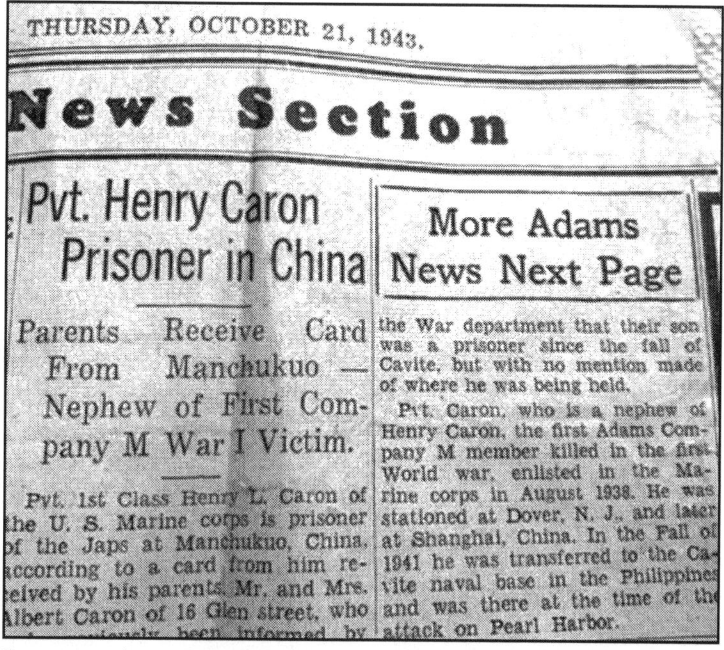

THURSDAY, OCTOBER 21, 1943.

News Section

Pvt. Henry Caron Prisoner in China

Parents Receive Card From Manchukuo — Nephew of First Company M War I Victim.

Pvt. 1st Class Henry L. Caron of the U. S. Marine corps is prisoner of the Japs at Manchukuo, China, according to a card from him received by his parents, Mr. and Mrs. Albert Caron of 16 Glen street, who ... been informed by the War department that their son was a prisoner since the fall of Cavite, but with no mention made of where he was being held.

Pvt. Caron, who is a nephew of Henry Caron, the first Adams Company M member killed in the first World war, enlisted in the Marine corps in August 1938. He was stationed at Dover, N. J., and later at Shanghai, China. In the Fall of 1941 he was transferred to the Cavite naval base in the Philippines and was there at the time of the attack on Pearl Harbor.

More Adams News Next Page

Transcript article announcing Henry L. Caron is a Japanese prisoner of war in Manchukuo, China. *[Courtesy of Murphy/Caron family]*

Henry arrived in Manchukuo, China, (Manchuria) in November 1942 and was incarcerated at Mukden POW camp located near the city of Hoten (now Shenyang). He was with approximately 1500 other American POWs that lived in frigid barracks partly built underground. Two hundred died the first winter.

For a brief period, Henry was assigned to work in the camp's machine shop, where he and his fellow workers did their best to turn out imperfect work to thwart the Japanese war effort (*Transcript* 10-27-1945).

For most of his time at Camp Mukden, he was assigned to one of its three separate factories where the men were used as slaves: Camp One was a tannery with 150 men, and Henry ended up there; Camp Two was a textile area; and Camp Three was a more dangerous steel and lumber mill.

In his homecoming interview with the *Transcript*, he related how his tannery comrades sabotaged production by skipping process steps and producing worthless leather. He was also candid about the unmerciful punishments meted out for sabotage and other lesser infractions.

He recalled, "A playful habit of the Japs when things of this kind happened (sabotage) was to force the entire group to strip and stand without clothes for hours at a time in an effort to have either the 'guilty' confess or someone 'squeal,' but it did not work. Another form of punishment would be to have a 'raised platform . . . . placed in front of the prisoners . . . four selected . . . .and beaten with a stick of wood two inches wide and two inches thick until they fell to the floor.'"

After he got home, Henry showed only his mom his disfigured, knotted, welted back—the result of public camp beatings. The physical and mental abuse would haunt him the rest of his life.

Despite the abuse, Henry remembered a Japanese physician, Dr. Oki at Mukden, "who used his own private funds to secure serums and medicines for the American prisoners and who did all he could to save lives and combat beriberi, scurvy, and dysentery that took the lives of so many" (*Transcript* 10-27-1945).

Few knew at the time, but the camp was also involved with Unit 731, the biological weapons group of the Japanese Army that conducted unspeakable lethal human experiments during the war. The unit purposefully infected prisoners with plagues and other diseases, conducted invasive surgeries removing healthy organs, and did unnecessary amputations, among a long list of atrocities.

Consequently, in addition to lack of food and medical care, these unspeakable medical experiments added to the camp's death rates.

Henry also mentions in his *Transcript* interview an underground black market in the camp that helped the starving men barter for food, often using cigarettes, canned items, or stolen leather from work. With winter weather approaching, amid temperatures of 40 below zero, long overcoats were a great cover, allowing men to smuggle items from their workplace to trade with others or local peasants.

Even with the help of the black market, food was little more than scraps of rotten vegetables and brown water. He continued to lose weight, but working in the tannery presented a small lifeline. Henry told the *Transcript*, "... upon the sight of a pig or cowhide that had a few scraps of flesh clinging to it, [I] would smuggle it, cut it up, boil it and make broth, if such a concoction could be called broth."

Almost a year after his capture, his parents received the following letter from the Marine Corps:

Headquarters US Marine Corps
Washington
March 15, 1943

My Dear Mr. and Mrs. Caron:

A partial list of American prisoners of the War in the Manila Bay area has been received from the International Red Cross, containing the name of your son, Private First-Class Henry L. Caron, confirming the fact that he is alive and a prisoner of War.

The report fails to state the place of his internment, but you may communicate with him by mail at the following address:

Private First-Class Henry L. Caron, U.S.M.C.
Serial Number 267082
Prisoner of War
Philippine Islands
c/o Japanese Red Cross
Tokyo, Japan
Via New York, N. Y.

It is requested that you address further inquiries regarding your son's welfare to the Prisoners of War Information Bureau, Office of the Provost Marshall General, War Department, Washington, DC.

Sincerely Yours,
C.P. Lancaster
Captain, US Marine Corps

Sometime after this terse communique, infrequent brief postcard notes were exchanged between Henry and his family. An undated one to his mother from "The Mukden Prisoner of War Camp, Mukden, Manchukuo" states:

> *Don't worry about me. I am in good health. I know this letter will find all of you in good health. Please have some of the other people write to me. I can receive packages. Inquire at the Red Cross. All of you keep smiling, I will be home. Your son. Henry L. Caron.*

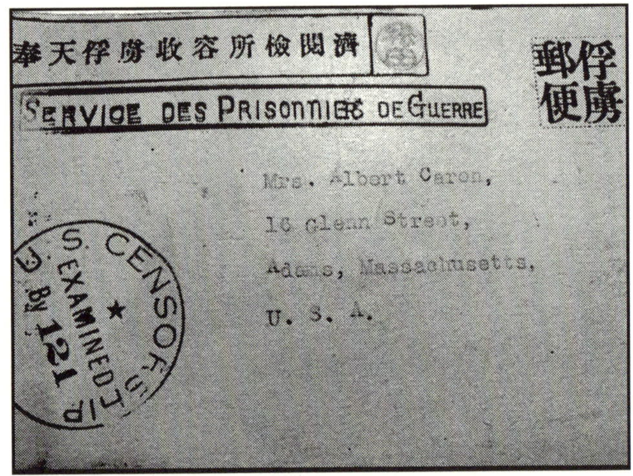

Postcard showing stamps by the censor and "Prisonniers de Guerre" mail service. *[Courtesy of Murphy/Caron family]*

An article in the *Transcript* on September 7, 1945, said Henry's family received a letter dated August 19, 1945, and in it he stated, "It will be the happiest day of my life when I reach home again." The *Transcript* noted, "There was no mention as to whether or not he had already been liberated."

Russia had declared war on Japan on August 8 and launched a large-scale invasion into Japanese-occupied Manchukuo, China, on August 9, 1945. The Soviet army arrived and liberated the camp on August 23, shortly after Henry had written his letter home. The Japanese had two POW camps in Manchuria, and Henry's camp was the largest with approximately 1300 remaining prisoners.

Several days after the Russians arrived, an American POW recovery team was on site to assist with the POWs' medical treatment, immunizations, processing forms, body recoveries for reburial, gathering of information on war crimes, and other preparations for the trip home.

On August 15, 1945, Emperor Hirohito announced Japan's unconditional surrender, which was formally signed on September 2, 1945.

Henry left Camp Hoten (Mukden) by ship for Korea, then travelled by hospital ship to Okinawa, then on to Guam for further hospitalization, landing finally on the US mainland in Oakland, California, on October 19, 1945.

Henry greets his family at the Adams railroad station upon his return home. (L-R: Father Albert, Henry, his mother Albina, and brother) *[Courtesy of Murphy/Caron family]*

He arrived home in Adams on Friday, October 26, 1945, for the first family reunion in over six years. He'd survived over three years as a prisoner of war in not one but two of the most infamous of camps. Not long after his return home, Henry married and moved south to sunny Florida, following his older brother Albert as he did so many years ago. Henry eventually owned a number of businesses, including a pecan farm and different amusement rides. Most of all, he enjoyed again being near his brother.

This Marine World War II hero is buried in Florida National Cemetery at Bushnell, Florida. His plaque reads:

> *Henry L. Caron*
> *Sgt. US Marine Corps*
> *World War II*
> *February 21 1920 – September 7 2000*
> *Prisoner of War*
> Semper fi

Henry's decorations include the POW Medal, the Asiatic Pacific Campaign Medal, the Victory World War II Medal, the American Defense Medal, and the Good Conduct Medal.

Between both Henry Carons, uncle and nephew, many medals were earned, much respect given. These two Berkshire boys, barely a generation apart, who never knew each other yet came of age just as their country needed them, voluntarily entered the two most fearsome wars of modern times. They showed what it meant to be heroic, tenacious, and hopeful until the end.

Staff Sergeant Rudolph Konieczny
10th Mountain Division
WORLD WAR II

The fighting, some hand-to-hand, had gone on all afternoon and his squad had cleared out a number of bunkers along the side of Mount Serra. At one point, Rudy took off by himself with his Browning Automatic Rifle (BAR) to clear out more fighting positions. As he climbed the ridge, he heard a noise and nestled down in some brush. Appearing before him were a number of Germans hurrying to reinforce their comrades below. Rudy quickly took them under fire, not realizing it was a squad of ten German soldiers. His BAR made quick work of those in the open, but there were several more ahead hidden in a dugout. He started out again, unexpectedly bringing himself directly into the firing line of a German MG 42 machine gun. Before he could react, the highly reliable gun stitched 7.92 caliber x 57 millimeter bullet holes across his chest and face down he fell, never to stand again.

Sempre Avanti (Always Forward) was the 10th Mountain Division's ski trooper credo in World War II. Among those vaunted men was Rudolph "Rudy" W. Konieczny, an Adams boy and championship skier, now known as the Hero of the Thunderbolt who gave his all to defeat the tyranny of Nazism.

RUDY'S STORY BEGINS WITH the immigration of his parents Karol and Zofia Szwajkos-Konieczny in the early twentieth century. Both were born in 1886 and immigrated separately through Ellis Island; Karol from Orlava, Czechoslovakia, and Zofia from Krakow, Poland. They were teenagers at the time.

Karol found work in Connecticut, while Zofia, after stopping in New York City, came to Adams, Massachusetts. Karol worked at an aluminum plant in Connecticut and at one point was incarcerated with other workers for assaulting their employer for back wages. While in jail, he heard a friend mention Zofia and upon his release he headed to Adams. The couple met, courted, and were married at Saint Stanislaus Church in 1911.

Karol and Zofia purchased a house on land that eventually became Cross Street. Karol added a second story to accommodate their growing family, as well as a barn, shed, and garage. The stuccoed, white brick house had four rooms downstairs, a kitchen, living room, and two larger entryways. Upstairs were five bedrooms and a bath. A cast iron wood/coal stove warmed the downstairs hallway. The home had no heat upstairs.

One story has Karol surreptitiously splicing into the electric box to wire the second floor. When the electric company caught him, his power was cut for a week as punishment. The house is now owned by Charlie Konieczny, a nephew, left to him and his sister by Rudy's brother Charles.

At one time, the 200-year-old house was a stopover for the stagecoach that ran through Adams. The east side of the house had an open cellar where the horses would be reined, while passengers and drivers debarked and went upstairs to eat and drink in the tavern.

Rudy was born at home on April 3, 1918, the seventh of nine children, four boys and five girls (a boy named Joseph died a few months after birth). It is believed that Karol, in some cases ineptly, helped deliver the children. As Rudy grew up, he shared a bedroom with his older brother Charles. For more room, they slept crossways on the bed with feet hanging over the side, often wearing jacket and gloves to fight the cold.

Initially, Karol worked in the local mills as a loom fixer and then for the New England Lime Company. Part of Karol's job for the lime company was breaking rocks using a sledgehammer and wedges. He grew quite skillful at reading the veins in rocks and would be called on to help his neighbors, showing them where to place the wedges to quickly break rocks on their property.

Rudy went to nearby Renfrew School from kindergarten to the sixth grade, dropping out to work at the Berkshire Fine Spinning Mills. Initially he served as a sweeper and an oiler, later probably as a loom fixer like his dad. During World War II, the mills wove cotton into material for military uniforms.

Rudy also helped on the Konieczny's small farm, tending his mother's garden and working with the cows and chickens. At one time, seven of the Konieczny's children worked at the mill. Zofia didn't want anyone to miss work, so if someone was sick she would go into work for them. She worked as an oiler, sweeper, loom fixer, and a doffer (one who removes full bobbins and replaces them with empty ones). Family members earned about $10.00 a week and on Fridays were paid in cash. After Zofia's room and board charge, the kids were given 25 cents as allowance. When Charlie's dad, Charles, turned twenty-one, his allowance went up to the grand sum of 50 cents!

Zofia also took in boarders, sponsoring those emigrating from Poland. Her boarders, up to eight at a time, usually spoke no English. They all resided in her 15 foot by 15 foot living room; in addition to a place to live, Zofia found them jobs at the mill. They were grateful and paid for their room and board, helping Zofia support her family of ten.

Early on, it was obvious that Rudy sought out adventure. Although small in stature at 5 feet 8 inches (compared to his father who was over 6 feet), he was wiry, daring, and had a charming boyish smile. His youthful appearance belied his age, and he was always open to a challenge. He and his best friend Greeny Guertin were daredevils. It is said that Greeny once scaled the steeple at Notre Dame Church just so he could wave to his friends. Rudy loved riding all over town on a bike without the benefit of brakes.

Rudy's antics, pranks, and sense of humor were well known in the little town of Adams. He was everybody's friend, occasionally a rule breaker, and while not appreciated by all, his daredevil actions usually drew smiles. Always willing to tease his sisters, when Zofia made soups or stews and Rudy wanted more to eat, he would finish his meal, walk around the table, and stick his finger in one of his sisters' bowls. They'd refuse to finish eating, so Rudy got extra.

He would also help his sisters when they came back late from a date by lowering a rope from the second story, hauling them back into the house. A stern Karol never figured out how they got around him.

When he built the garage, Karol had found many copper Indian Head pennies, possibly some valuable from the stagecoach era, and collected them in a jar. Rudy, not understanding their potential worth, found and spent the pennies on candy. When Karol found out he tried to buy them back from the store owner, to no avail.

As a teen, Rudy bought an old Packard with a stick shift. He let his dad, who did not drive, try the car out. Rudy had playfully placed the transmission in reverse so when his dad got in and popped the clutch he immediately went across the road and hit an embankment. They had to "rope" him out of the hillside.

As a youth, Rudy played football with the Saint Stanislaus team and had a very brief boxing career, appearing at the Meadowbrook Arena in North Adams coached by his brother Charles. In Rudy's only match, he was apparently knocked out in the first round and that ended his boxing career. As he got older, Rudy, smaller than Charles, used to go to bars for a few beers. There were times when Rudy would pick out the biggest guy in the bar and initiate a scuffle, knowing Charles would help him out. After one such scuffle, Charles told Rudy, "I'm not doing that anymore."

He often tagged along with Charles, who was six years older and an accomplished local athlete. Rudy looked up to him and knew he couldn't compete with him in the typical strength or speed sports. Charles did not like skiing, however, and Rudy saw an opportunity to excel on his own.

In 1934, the Savoy, Massachusetts, Civilian Conservation Corps started a project that would change Rudy's life. Thirty young men from the C's spent four months carving a two-mile expert ski trail on the steep east slope of Mount Greylock, to be called the Thunderbolt. Rudy had left school at an early age to work in the mills; his expectations were modest. He had started skiing at fifteen years old (in 1933), and now one of New England's biggest ski challenges was literally in his backyard. (For more detail, the creation of the Thunderbolt ski trail is enjoyably described in a Hoosac Valley documentary called "Purple Mountain Majesty.")

Once the Thunderbolt was completed, Rudy would often hike up the mountain after work with his skis on his back and then ski all the way home to Cross Street. During his initial downhill runs, with no experience at stopping, he and Greeny would just flop into a snowbank. At the time, there were no other houses in the area, and he would ski directly to his house, arriving with icicles hanging from his nose and ears and frozen eyebrows. He might take a few moments to have some of his mom's soup, then turn around and head right back up the mountain, sometimes making the trip three times in a day. After many of these forays, he decided to save his money during the year and quit working at the mill during the snow season so he could ski all the time.

The young skier looking dapper in knit earmuffs, necktie, and light parka. *[Courtesy of Blair Mahar and Konieczny family]*

He told his sister that he would leave his lunch on top of the mountain just so he would need to go back up one more time. Rudy, with his strong desire to compete, had his own rigorous training schedule, often accompanied up the mountain by friends who would someday join him in the Army's 10th Mountain Division. His brother and others said he displayed absolutely "no fear cutting corners on the trail" when negotiating the treacherous, steep two-mile run. He won his first race at age sixteen.

Racing was unsophisticated in the late 1930s. Everyone carried their skis on their backs, taking more than half an hour to climb the mountain. If they were packing down the trail on top of fresh powder, it took at least two hours. At the top of the mountain, a bottle of ink was used to etch the starting line in the snow. The tips of your skis had to be aligned with the black line, and a primitive wire system would chortle out buzzes. At the third buzz, the starter would let go of the skier's pants and the race was on.

In early 1938, in time trials sponsored by the Thunderbolt ski club, Rudy skied the two-mile course in less than three minutes. Later in the year, in the Massachusetts Downhill Championship, Rudy, the nineteen-year-old farm boy, helped win the team trophy with a time of two minutes and fifty-four seconds.

One disappointing event in February was the Eastern Championship loss to some very accomplished German skiers, one of whom shattered the course record held by Rudy. Although skiing with a sprained ankle, Rudy still placed sixteenth among a cast of worldwide skiers. It is notable that the European "winds of war" were starting to appear with Germany's annexation of Austria and Czechoslovakia. It would not be too long before some of these Adams skiers could be meeting their German counterparts on the field of battle.

In the same year, representing the Adams Thunderbolt Club, Rudy competed in the National Downhill Championship Race in Stowe, Vermont, losing when he failed to go through a gate, although gathering considerable experience and meeting other world-class skiers.

Rudy and the Ski Runners competed a number of times in the Massachusetts State Downhill Ski Championships against top-notch competition from Dartmouth College, Lake Placid Ski Club, and others. In another race, the Eastern Downhill ski championship, Rudy was recorded making the run in two minutes and twenty seconds, helping the Adams team place second overall.

Rudy finishing a run on his beloved Thunderbolt. *[Courtesy of Blair Mahar and Konieczny family]*

Over years of racing, Rudy improved his Thunderbolt time to just two minutes and eight seconds. During the years of downhill skiing, most of the young men worked in the mills earning very little money, which complicated their ability to buy good ski equipment. A local furniture owner, Art Simmons, befriended them, creating a ski section in his furniture store and selling equipment on credit to the local skiers. It is said his terms were quite generous.

Skiers of the era often used the popular Griswold hickory skis, and the Thunderbolt skiers from Adams presumably bought them from Art. (Interestingly, the 10th Mountain issued Griswold skis to its troopers during their training.) It is said that some of the local skiers would take advantage of the ski company's one-year guarantee and after a hard winter's use might "accidentally" break a ski tip and receive a new pair.

Few people knew that Rudy loved to sing; he became an ardent yodeler while traversing Mount Greylock's slopes. He was loud and boisterous, yet patient enough to teach his younger brother Adolph to yodel.

In 1940, Rudy and Greeny, his best friend, were both admitted to the same room at the Adams hospital with broken legs. Even there, they continued to be pranksters and tested the rules by smoking. Whenever a nurse came by to check on them, Rudy or Greeny would yell "load the cannon" and while one of them would extinguish the cigarettes, the other would place the extinguished cigarettes inside an extra toilet paper roll hidden on the floor, stuffing the ends with tissue to seal off any smell. The staff was relieved when the two miscreants were finally released.

In autumn 1940, with war on the horizon, Rudy took the unusual step of joining the Army's Coast Artillery. Once war was declared, after the Japanese attack on December 7, 1941, there came a demand to create an alpine ski unit. After securing three letters of recommendation, in 1942 Rudy and a friend were headed to Fort Lewis in Washington State and the slopes of nearby Mount Rainier. (Later, as qualifications lessened to join the 10th Mountain Division, senior skiers, like Rudy, were nicknamed "the three-letter men.")

During the following months, approximately twenty Adams men volunteered for the 87th Regiment of the 10th Mountain Division. This was considered one of the largest groups of volunteer skiers from one town, and Adams had just 12,000 citizens.

Rudy, already an accomplished skier, now had the opportunity to train others. Never having been out West, he was able to ski in some of the most beautiful areas of the country. His confidence and leadership skills grew accordingly, and on weekends he was able to compete in races on the 14,000-foot Rainier.

For the next year, training was arduous and grueling. The men trained in all types of foul weather, wearing skis or snowshoes and carrying rucksacks full of equipment, estimated to weigh 90 pounds. This did not include their M-1 rifles and ammunition. Rudy often carried his squad's BAR, which weighed twenty pounds.

When not seriously training, Rudy continued to be the prankster. As one story goes, he was demoted from the rank of Master Sergeant to Tech Sergeant when he disrupted a boring class by leaving the room, climbing on the roof, and putting a blanket over the chimney. Smoke poured into the classroom, ending that course of instruction.

There seemed to be a number of times when he appeared in Canadian or Italian officer uniforms, always up to some shenanigans. There was also believed to exist a video of him dressed as a priest, performing some "official" rite. Of course, weekends skiing and drinking with buddies were prime entertainment.

Rudy, on the right, training out west before deployment. Notice his mess kit hanging in the tree over his shoulder. *[Courtesy of Blair Mahar and Konieczny family]*

At one point, Rudy, due to his expertise, was selected to test new equipment with a group of soldiers in the Canadian Rockies and then head to Camp Carson, Colorado, for more training. Once again, he found wonderful skiing conditions at nearby Pikes Peak.

The 87th Regiment's next stop was Camp Hale, a new training center in Colorado that would accommodate thousands of inductees. Again, the skiers would escape from base on the weekends to ski in a nearby old mining town called Aspen, where the powdered snow was perfect.

It wasn't long before the troops left Camp Hale for Fort Ord in California, then on to Adak, Alaska, preparing for Operation Cottage, the battle for Kiska Island, which was occupied by Japanese Marines. The invasion involved a number of units including the 87th Mountain Infantry.

In August 1943, Rudy's unit landed on Kiska in miserable conditions and discovered after moving inland that the Japanese had abandoned the island. That did not prevent a number of casualties from friendly fire with Canadian troops, enemy minefields, booby traps, and accidents. In addition to damaging a nearby ship, the Allies suffered over 300 dead and wounded.

Chastened by the experience, the 87th returned to Camp Hale for more training. Rudy continued in his well-respected role as a leader and a trainer. In the summer of 1944, the unit debarked for Camp Swift in Austin, Texas. Moving from the Rockies to this hot, sandy outpost awaiting further orders must have been frustrating to mountain troopers. They were hearing the news of the war raging in Europe and the Pacific while they endured seemingly endless marches in an arid climate.

They did not have to wait long. The 10th Mountain Division was moved to Hampton Roads, Virginia, in December 1944 in preparation for crossing the Atlantic to some secret destination. Unknown to them, General Mark Clark had decided the Mountain troops were just what he needed to fight the Germans who still remained in northern Italy along the Apennine Mountains.

Arriving in Italy in mid-January 1945, the Mountaineers were ferried quickly by train to the front lines. No one realized the war would be over in less than four months. The unit began with patrolling, soon realizing they were up against battle-hardened, dug-in German Alpine divisions. Realistically, these could be men Rudy skied against years ago.

An experienced squad leader, well respected by his men, Rudy assumed more responsibility, watching over and worrying about his men. He was always out front with his BAR, often volunteering for dangerous night patrols or going out by himself to eliminate bothersome snipers.

The 87th was now in the Apennine range north of Pisa, with directions to seize Mount Belvedere, although realizing the nearby Riva Ridge that overlooked Belvedere would need to be captured first. In a daring nighttime raid, Rudy's battalion successfully seized the Ridge, and the 10th then overran Belvedere. Often fighting uphill against dug-in-troops, Rudy and his squad were frequently pinned down by enemy fire and grenades.

At one point Rudy used two BARs on bipods to relieve the pressure on his men pinned in the open. Fighting was fierce and hand-to-hand; few prisoners were taken.

On March 3, 1945, Fox Company attacked a low hill called Malandrone on the way to their objective of capturing Castel D'Aiano. Under heavy shelling, while charging uphill, Rudy was hit, shrapnel piercing his arms, side, and legs. Evacuated to Livorno, then to hospital in Naples, Rudy was out of action. The doctors told him he was headed home.

Worried about his troops and missing the action, Rudy left the hospital without permission the week of April 2, 1945, returning to the front lines. It was just in time for the 10th Mountain Division's spring offensive on course to take control of the Po Valley.

On April 14 and 15, the 10th engaged in a series of bloody battles taking several hills and nearby towns. The dogged determination of the 10th's troopers resulted in many casualties.

On April 16, Rudy's depleted squad led the way to take the town of Torre. The vicious fighting continued into the next day, and his unit was told to attack a small ridge called Mount Serra. The fighting devolved into hand-to-hand, and it was most vicious. Rudy, with his

BAR, went off by himself to clear more enemy bunkers. He was not seen alive again.

The Konieczny family received a Missing-in-Action telegram in early May, two days after the war ended in Europe. Another telegram arrived on May 21, 1945, telling them Rudy had been killed in action on April 17, 1945. His body was found in thick foliage at the base of Mount Serra.

Larger than life, Rudy, the local skiing superstar, prankster, everybody's friend, leading from the front, died shortly after his twenty-seventh birthday and three weeks before the war in Europe would end. Fighting in some of the roughest terrain of the war, the 10th Mountain Division in a matter of four months of combat lost 992 killed, with over 4000 wounded soldiers.

For a while, Rudy's legacy was kept alive by ski races on the Thunderbolt dedicated to him. A warming hut at the top of Mount Greylock was also dedicated to Rudy. Most recently, a banner with Rudy's image and unit was erected in front of the First Congregational Church in Adams.

After four years, Rudy returned to Adams. His mom, dad, niece, and many townspeople met the train that brought his body back. On March 19, 1949, he was interred at Bellevue Cemetery with military honors. A man many thought destined to be an Olympic skier now rests peacefully in a plot with a view of his beloved Thunderbolt Trail.

Postscript: *During Rudy's assignment out West, he and his friends discussed plans to purchase land on which to build ski resorts in this pristine winter area. Those who returned from the war did just that, developing trails, teaching skiing to youths, and promoting the sport extensively. Rudy, the medium-height, fair-haired, baby-faced Tech Sergeant, would have been proud.*

LT. COMMANDER LEONARD KOCZELA & LIEUTENANT RUTH KOCZELA
LANDING CRAFT COMMANDER & CODE GIRL
WORLD WAR II

Ruth and Paul Koczela, Naval officers *[Courtesy of Koczela family]*

As an ensign, Ruth Black led a group of young naval women who were part of a vast effort to decode enemy messages during World War II. From all walks of life, these young women moved to Washington, DC, by the thousands and successfully broke Japanese and Nazi ciphers.

After the war, members of Congress and leaders of the Armed Forces attributed the intelligence gathered primarily by these female code-breakers to "shortening the war by no less than two years" and "saving many thousands of lives."

Simultaneously, Ruth's future husband, Leonard "Paul" Koczela, also a naval officer, was thousands of miles away navigating a landing craft under enemy fire. Paul was delivering Marines who would wrest highly contested Pacific atolls away from the enemy.

Paul was born and raised in Adams, Massachusetts. He met Ruth at the State Teachers College of North Adams. They graduated together in 1942 and began their astounding adventure.

Ruth Evelyn Black

RUTH WAS BORN ON June 25, 1921, in the tiny hamlet of Searsville within the town of Williamsburg, Massachusetts. She celebrated her 101st birthday in 2022. Her father was John Milton Black, a subsistence farmer, and her mom was Alta Estella Bates, a homemaker. The family of three girls and one boy grew up in the same house in which their father John was born and raised. His father had been Lewis Train Black, who fought for the Union Army in the Civil War.

During the Depression, they always had enough to eat. The family planted, tended their crops, and managed their livestock. They had six milking cows, horses for plowing, chickens for laying, and pigs for meat and lard. Ruth, the youngest of four, weeded, took care of chickens, collected eggs, and sold raspberries and blueberries at their roadside stand. The children were all born five years apart, so Ruth's brother Lewis was 15 years older than she was.

As a young girl, she loved the outdoors and could identify most birds. She enjoyed trout fishing with worms and hunted with her own .22 rifle. She was also an avid Girl Scout.

Her dad was known for his poetry and was called "the Bard of Searsville." He was known to always carry a pencil stub and slip of paper on which to write poems, or he wrote them on the barn walls or stalls, some still in evidence today. A book, *Homespun* by John Black, collected over 100 of his poems.

Ruth attended Williamsburg Elementary from first grade through high school, self-admittedly more interested in sports than studies. She loved baseball and excelled at running; she has the medals to prove it. She also was good at basketball. During this time, she learned the intricacies of playing the harmonica, and "Red Wing" became one of her favorite songs.

In addition to her farm chores, she worked as a waitress at the nearby Snow Farm in their small inn. With her parents' encouragement, she and her best friend applied and were accepted at the State Teachers College of North Adams, about forty minutes from home, and they began classes in 1938.

During orientation and on-stage introductions, she noticed Leonard Koczela, nicknamed Paul, a day student from nearby Adams. He had seen her during the same time period. Shortly afterward, they began dating. Ruth noted later that only five men were present; many had already gone into the military.

While both fathers liked Ruth and Paul, they were adamantly opposed to any serious relationship. The possibility of Ruth, a Protestant farm girl, marrying Paul, the son of a Catholic baker, would never work. The young couple kept dating and did not discuss the issue with their dads.

Leonard Stanley Koczela

LEONARD WAS BORN IN Adams in 1918 to Ignacy "Ernest" Koczela and Ludwika "Louise" Fryc. Both emigrated from Poland via Ellis Island in the early 1900s, settled in Adams, and married.

Ernest, a baker, operated "Homelike Bakery" on Hoosac Street for many years, and during this time the couple had Leonard, three more sons, and a daughter. (After Louise passed away unexpectedly, Ernest married Mary Baldyga, and their family grew to nine children.) Ernest also managed the Polish Co-Operative and Hammond Bakeries. Most of their children worked at the bakeries before or after school, including Paul. His brother Thaddeus operated the well-known Ted's Bakery in nearby North Adams.

Leonard received the nickname *Paul* early in life when his older brothers drafted him for family boxing matches. He usually ended up with a bloody nose, similar to a famous boxer at the time, the Great Paolino—hence, Paul.

Paul attended Saint Stanislaus grammar school. His family struggled during the Great Depression and had little to eat. He remembers having a cup of milk to drink each day, some stale bread leftover from the bakery, and, unsurprisingly, he often went to bed hungry. Paul worked at the bakery before school delivering goods and then returned after school, working until ten p.m. He still managed to skip the second and sixth grades. He continued to excel while at Adams High and graduated at the top of his class in 1938.

After he and Ruth started dating in their first year of college, they enjoyed college activities together and thrived academically. They were both involved in the Drama Club; Paul was the sports editor, belonged to the Newman Club, and served as class treasurer for one year; Ruth was junior class president and student council president.

Paul borrowed a car during their college courting days and visited Ruth at her farm. They only lived about thirty miles apart. He and some friends dropped by one day just when her mom had told her to go kill and pluck a chicken for dinner. Ruth had the chicken tied up outside and was getting ready to chop off its head when Paul and his friends arrived. Feeling unladylike, she hid the hatchet behind her back. At about that same time, her mother came outside and yelled out loud: "Ruth, that chicken won't kill itself."

Ruth was embarrassed. Paul's interest was not diminished.

Upon graduation, their paths diverged; within the throes of World War II, Paul joined the Navy and left for Officer Training School at Notre Dame University. For one year, Ruth accepted a job teaching grades four through six at a one-room schoolhouse in Monroe Bridge, Massachusetts.

During the summer of 1942, while Ruth was preparing to teach, President Franklin Delano Roosevelt signed a law that created the women's naval reserve known as the WAVES (Women Accepted for Voluntary Emergency Service). The purpose was to free up men for sea duty.

Smith College in Northampton, Massachusetts, was transformed into the "USS *Northampton*" and became a training base for female naval officers. The Alumnae House would be HQ for the United States Naval Reserve Midshipmen's School. Bunk beds were installed in rooms. Dining halls were converted to classrooms. Training was called not *boot camp* but *indoctrination*. Women were to arrive as seamen and, upon completing the training, be commissioned as ensigns.

There was great pride in joining the WAVES. During the war, women from all walks of life, even those with college degrees, wanted to serve. A sense of patriotism arose, a hope of adventure, and a desire to see if they could pass the entrance test. The smart uniforms didn't hurt.

Ruth applied for the WAVES, but was rejected because of a mistake regarding her height. Ruth finished her school year and applied again, along with her best friend. She requested, and received, a letter of recommendation from the president of her alma mater, the State Teachers College of North Adams. It arrived on letterhead stationery. This time Ruth was accepted, and in 1943 she began Midshipman School at USS *Northampton*.

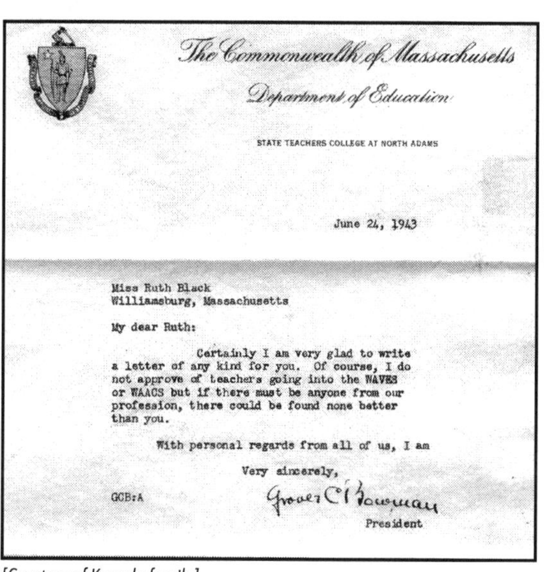
The Commonwealth of Massachusetts
Department of Education
STATE TEACHERS COLLEGE AT NORTH ADAMS

June 24, 1943

Miss Ruth Black
Williamsburg, Massachusetts

My dear Ruth:

Certainly I am very glad to write a letter of any kind for you. Of course, I do not approve of teachers going into the WAVES or WAACS but if there must be anyone from our profession, there could be found none better than you.

With personal regards from all of us, I am

Very sincerely,

GCB:A

Grover C. Bowman
President

[Courtesy of Koczela family]

She came to love her time at "Hamp." Her letters home recount serving as a platoon leader, enjoying the great food, learning calisthenics, marching to classes, and making new friendships. Early morning reveille, room cleanup, making her bed, and marching to breakfast at the "Hotel Northampton" were all part of the fun. She was also on the basketball team. This small-town farm girl adapted well to the Navy.

Meanwhile, Paul was commissioned as an ensign in February 1943, then attended Amphibian Warfare Training School in Virginia. He had further training with the Marines at Ft. Pierce in Florida. In August 1943, Paul was assigned to the USS *Elmore*, an attack transport ship with a crew of 600 and the capability of bringing well over a thousand troops to war. The *Elmore* carried twelve LCVPs (Landing Craft Vehicle Personnel) that could ferry invasionary troops to shore. Paul would be the commander of one of these crafts.

Once the ship was refitted in December 1943, it started for the Pacific with 1600 Marines on board from 4th Division. Early the morning of January 31, 1944, Paul began navigating his boat in the first wave to land Marines onto the beaches of Roi-Namur of the Kwajalein Islands, as part of Operation Flintlock. He and the crew searched for the preestablished landmarks on which they had been briefed, but bombers had obliterated them during the night. Using his nautical intuition, he directed the ship to forge ahead, providing a continuous shuttle of men and supplies for the invasionary force and evacuating the wounded for six days under fire.

It was a horrible time for most of the sailors, bringing back the dead and wounded Marines, some of whom they knew (and others whom they had just brought ashore). Paul remembers returning to the beach with fresh troops and finding those previously dropped off floating in the water dead. Paul's bunkmate, a Marine captain, was killed shortly after landing when he triggered a land mine. Paul seldom talked about the landings.

After the battle, the USS *Elmore* spent months refitting and resupplying, getting ready for the next invasion. During this time period, Paul was promoted to Lieutenant Junior Grade. His boat was assigned yet again to the first assault wave in Guam, landing near Apra Harbor. Boat losses were fierce, the harbor and beaches had been mined, and he and his crew saw many being blown clear out of the water. After three days and nights of unloading troops and resupply, they departed without serious injury, although Paul developed acute headaches and sleepless nights and eventually was transferred stateside.

While Paul had been in the Pacific, Ruth had been assigned to the US Naval Communications Annex in Washington, DC. The Navy had been looking for college-educated women who were bright, intelligent, resourceful, had good memories, and were also patriotic and unattached (not married or engaged). Former schoolteachers seemed to be well suited, as did librarians, in organizing and analyzing data.

Before the war was over, there would be thousands of female officers assigned to Army and Navy Operations in Washington, DC, with the sole purpose of breaking and deciphering enemy-coded transmissions. The Navy took most of the responsibility for breaking Japanese codes; the Army worked on deciphering German transmissions.

These specialists were often divided into small groups and would come to be known as "code girls." Working in signals intelligence, acting as cryptanalysts, they deciphered Japanese and German codes, ran complex machines, tested our own codes, and even helped create dummy traffic to fool the enemy.

Before Ruth's arrival, one of the unit's greatest triumphs was the intelligence and advantage it provided Admiral Nimitz for the Battle of Midway, which many have called the turning point in World War II.

Ruth was at the center of these continued efforts, leading a section of code girls in successfully deciphering Japanese war correspondence. The Annex, as it was known, employed thousands of WAVES during wartime.

The Annex was formerly the Mount Vernon School for Girls before being commandeered by the government. (It now houses the Department of Homeland Security.) Ruth lived in the Woman's Officer Quarters Barracks D (Room 202), right across the road from the Annex. It was the largest women's barracks in the world, housing almost all codebreakers. On one special occasion, Ruth met Eleanor Roosevelt when she came to their quarters to speak to the WAVES.

Upon coming aboard the Annex, each WAVE signed a secrecy oath swearing not to ever discuss her activities under penalty of prosecution under the Espionage Act. For the rest of her life, Ruth took this oath very seriously and spoke little of "her girls' activities." One time in DC, around the time she was taking the Oath of Secrecy, she got a ride to WAVES Headquarters, and the driver asked her what she did. She said secretarial work and never volunteered where or for whom she worked even after continued questioning. She later found out the driver had been a security plant to see if the WAVES would reveal what they did.

Ruth was known for her management efficiency and ability to organize workflow. She directed the efforts of ten to twenty enlisted navy women. They were a close-knit group, spending long hours deciphering intercepted Japanese military messages. Some were passed to them from England's MI-6 Intelligence Unit in Bletchley. Each woman would decipher a small portion of the intercept and pass it on to the next person. Remarkably, no one in Ruth's unit spoke Japanese; they would look for repeated symbols, numbers, or word patterns.

Codebreaking involved many teams coordinating their efforts looking for some common thread, something they remembered seeing in another message, maybe recalling a coincidence, intuiting something, or following a hunch. A certain persistence or doggedness was an excellent quality to possess.

If a message was uncovered, it was passed on to Ruth, and she quickly brought it to the officer-in-charge. The "girls" would notice when a large military operation was about to commence; their activity would increase. They would learn to connect a subsequent battle in the Pacific with their intelligence-gathering efforts.

The job required attention to detail and a high degree of concentration, but it wasn't without its lighter moments. One time during an idle period, Ruth told her "girls" a message was coming in, and one of them responded, "But, Miss Black, my nails aren't dry yet," having just applied a coat of nail polish.

Ruth's letters home during this time have been preserved and reveal a hometown girl enjoying newfound friends, trips to various cities, sightseeing, watching the latest movies, arranging for her mom to visit, counseling the enlisted women who worked for her, discussing the future and Paul, seeing the New York City Rockettes, and searching for a flat with girlfriends. More seriously, she also notes how polio was rampant in one school district.

In her letters, Ruth also mentions receiving care packages from her mom with cookies, jam, and the clothes she'd requested. She and her friends enjoyed bowling, movies, plays, theater, horseback riding, football games, touring the Smithsonian Institute and local monuments, and visiting New York City and Williamsburg, Virginia. Ruth also mentions several times how much she and her friends liked their uniforms and enjoyed marching in Memorial Day parades.

She notes the majesty of the National Cathedral, that it was still under construction, and cites her infrequent visits from Paul. Nothing is ever mentioned about the type of work she was doing.

She and a friend were allowed to take free "hops" (flights) from the nearby Naval Air Station just for something different from their routine trips, mostly on cargo planes. Once, unknowingly, they got on a one-way trip to a Nebraska airplane graveyard and had to ride in a Jeep all the way back to DC.

Time passed quickly for Ruth, and she started seeing more of Paul when he was reassigned as a navigator to the USS *Vanadis* based in Newport, Rhode Island. After a short period of time, as the war was winding down, the *Vanadis* was decommissioned, and Paul was assigned to the Hydrographic Office as a cartographer in Maryland, close to Washington.

For a number of years Paul had carried around an engagement ring to give to Ruth, and if not for war, long separations, and their fathers' combined disapproval, they might have been wed right after college. In a bittersweet turn, when Ruth's dad passed in 1943 and

Two happy young naval officers tie the knot. *[Courtesy of Koczela family]*

then Paul's in 1945, the young couple became engaged and were married in 1946 at the US Naval Chapel at the Communication Annex. They were both honorably discharged from the Navy shortly afterward.

Ruth was released as a Lieutenant, and her decorations included the Navy Unit Commendation, American Campaign Medal, and WWII Victory Medal. Paul, who stayed in the Reserves slightly longer, was released as a Lieutenant Commander and received the American Campaign Medal, the Asiatic Pacific Medal with seven battle stars, and the World War II Victory Medal.

Even though Ruth and Paul were thousands of miles apart during the war, the efforts of Ruth and her code girls directly supported and significantly impacted Paul's theater of operations, the Pacific Ocean.

The codebreakers were credited, in the end, with sinking fifty of the Japanese tankers, freighters, or cargo ships that resupplied the islands where Paul's invasionary forces were attacking, often denying enemy soldiers needed ammunition, food, and supplies. Intercepted codes also enabled US naval forces to sink Japanese and German submarines, direct threats to US ships. At the end of the war, code girls were reading messages almost as fast as they were generated, allowing near-simultaneous reaction.

Life Together

After the war, Ruth adapted and became a homemaker. The family quickly grew to five boys and a girl. Later on, she turned to substitute teaching and learned bookbinding skills. Paul became an analyst and cryptologist with NSA (Naval Security Section) at Fort Meade, Maryland, and spoke little about his work. His love of mathematics, foreign languages, and experience with cartography served him well in his role with the intelligence agency.

He traveled to Russia and Turkey during several tumultuous periods and spent almost thirty years working for the government. One of his trips might have been connected to the "U-2 Incident" of 1960, in which the reconnaissance plane of American agent Gary Powers was shot down over Russia, but no one is sure.

Growing up, the children were advised when completing school forms that dad's occupation was "security" or "an analyst." Unlike most families, the dinner-table conversation revolved around mom's and kids' activities, not dad's. Even today, his children only presume his work must have involved assessing satellite information. One of his children bought Paul *The Puzzle Palace*, a book about the inner workings of the NSA, and he never read one page. Presumably, Paul considered it disloyal for someone to write about their secret work.

Vacations were wide-ranging for a family of eight. The packed Ford station wagon could be found traveling from Mammoth Cave and Abe Lincoln's Cabin in Kentucky to Niagara Falls, or across the country to Yellowstone National Park. The wagon took them camping and to the Florida Everglades. The family, or sometimes just the kids, would go back to work on the family farm in Williamsburg. At one point, they bought their uncle's nearby farm and vacationed there for many years.

During the war, Paul had accumulated a few personal effects of some Japanese soldiers. As time passed and he continued to think about the effects of the war, he thoughtfully decided to return the property to the Japanese embassy, in hopes that they could locate the owners' families.

Paul has been described as quiet yet social, reflective, a hard worker, a puzzle solver, dedicated to the family, and serious about his work. His past experiences as a baker would surface for the holidays, and he liked to bake Polish *bulka* raisin bread. Few people knew he was also a cake decorator. He saved the neighbor's wedding cake when it was being delivered next door on a warm day, and some tiers started to slide off. He told the van driver to bring the cake to his house. He opened the van's side door and quickly repaired the cake as only a former baker could, saving the day.

Though an active member of the Phoenix Society, a social club for NSA retirees, he never did discuss his work with the family. Rarely did he discuss his time in the Pacific during World War II, although his family knew the memories bothered him, and some presumed he suffered from PTSD. Over the years, he and Ruth accumulated many friends who enjoyed playing bridge and traveling. Paul died in 2003, and his ashes are interred at the Arlington National Cemetery.

Ruth was always busy with the family's day-to-day activities. The kids often came home for lunch; preparing meals and taking care of a family of eight was a full-time job. As a mother, she preferred being outside, had a great sense of subtle humor, and was the family's first line of discipline. Even now she is talkative, and few can get things past her.

Ruth has remained active her whole life. At 101, she has her own

At age 100, Ruth is honored by the Navy's Ceremonial Guard and her family in the very chapel in which she was married 75 years earlier. (From left, Julie Koczela, Jack Koczela, Seaman Josephine Rojas of the US Navy Ceremonial Guard, Ruth Koczela, Lt. Cmdr. Julie Gillespy of Naval District Washington, and Luke Koczela.) *[Courtesy of Koczela family.]*

ground-floor apartment in her son's house, maintains her sharp wit, and can muster one-line retorts that bring smiles. She still enjoys gardening and not long ago had to dispatch a voracious porcupine that was raiding her raspberries.

Although released from the secrecy oath she took as a codebreaker almost eighty years ago, she still takes it seriously and talks very little about those times. Some years ago, she was introduced as "a code girl" by author Liza Mundy to the group at a book signing. What wasn't said was that Ruth had declined to be interviewed for the book, although several of the women she worked with participated.

Never one to remain idle, at eighty-nine years old, Ruth skydived from the nearby Northampton airport. (She says President George H. W. Bush gave her the idea.) In her nineties, she traveled to Kenya with two of her sons to go on a safari that included a hot-air balloon ride. She also has zip-lined and rafted on the Connecticut River. Ruth still plays her harmonica whenever she gets a chance.

Surely, both fathers would be proud of the lives led by a Protestant farmer's daughter and a Polish Catholic baker's son: These two young kids from the Berkshires celebrated a long marriage and became accomplished naval officers, changing the fate of the world two people at a time.

Lt. Colonel Sterling S. Burnette

Distinguished Service Cross

World War II

A young man from Adams, Massachusetts, recently battlefield-promoted to a battalion commander, finds himself almost 4000 miles from home in charge of capturing a critical bridge crossing in a small French town. The bridge is mined with explosives and covered by German weaponry. Lt. Col. Sterling Burnette rallies a small group of soldiers and races with them for the bridge. Almost immediately, they come under intense enemy fire.

Members of the group successfully reach the other side, securing the bridge; Sterling is not one of them.

STERLING "CURLY" STURTEVANT BURNETTE'S parents, Grover C. Burnette and Florence Sturtevant, were born just months apart in Savoy, Massachusetts, in 1884. Presumedly, their families knew each other and socialized together in this town of fewer than two hundred families.

Grover and Florence were married in 1906 and moved to Adams, Massachusetts, in 1916. The family moved to the Zylonite section of town (so named after a company that manufactured cellulose nitrate products). The couple rented several apartments, eventually ending up on Columbia Street near the lime quarry, where Grover found work as a laborer.

Tragedy struck the young family when Grover died at home at age thirty-four from pneumonia attributable to the 1918 influenza pandemic. He was one of 56 Adams residents who succumbed to the disease. Florence was now a single mom with no source of income, three small children, and five months pregnant with a fourth. She never remarried and raised the children by herself.

Called "Ma" by everyone (and widely known as a resolute personality), she supported her family as a seamstress, working long hours for a millinery shop, stitching, and in some cases, knitting and crocheting while also maintaining a large garden to bolster her food supply.

Sterling Sturtevant Burnette was born July 26, 1913, the third of Florence's children. She gave him her maiden name as his middle name. Intriguingly, Sterling's birth certificate lists his last name as Burnett with no e at its end. Few people know the story that lies behind this curiosity. At one point, some Burnett family members offered to take Sterling off Ma's hands and raise him. The "offer" offended Ma, and she decided to differentiate her family from the rest of the Burnetts, adding an **e** to their names.

Early on, Sterling's wavy hair earned him the nickname *Curly*, which remained with him throughout his life. He attended Howland Avenue School from kindergarten through eighth grade and, at thirteen, was one of the first members of the newly formed Zylonite Scout Troop 4. Sterling was proud of his role as the assistant patrol leader for the group, also called the Panthers. They met every Wednesday night at the Zylonite Community House.

He attended Adams High School on Liberty Street, where he played on the football team for three years and was chosen All-Berkshire guard in his senior year. He was also on the senior honor roll and a yearbook staff member.

On June 19, 1931, the senior class held the Annual Class Day exercises reflecting and celebrating the class's accomplishments and developments over the last four years. Curly would serve as the class "prophet" that Friday night, delivering a humorous prophecy foretelling what the forthcoming years would have in store for each of the young graduates, including himself, little realizing the World War that was on the horizon, or how it would dramatically affect everyone's lives.

After high school, in the 1930s, Curly was busy joining a local semi-pro football team called the "Zylonite Eleven"; enlisting in local Company M of the National Guard; and getting a job at General Electric in Pittsfield.

Curly in his football uniform, ready to take the field. *[Courtesy of Tarsa/Burnette family]*

Curly played for years, through the 1930s, on the formidable Zylonite Eleven, a frequent contender for the Western Massachusetts semi-pro football title. The team, a local favorite, played throughout the eastern states, including a game at the Sing Sing prison in Ossining, New York, against a team called the Black Sheep. Injuries were frequent and substitutions commonplace, with only rudimentary leather helmets and pads for protection. While Curly played guard at times, this medium-built player was also pressed into action as the team's center against some heavyweight players.

In 1933, Curly joined Company M, 104th Infantry of the Massachusetts National Guard as a private and worked diligently in Company M's administration. He was quickly promoted to corporal and sergeant. After several years of drills, training, and study, he passed the lieutenant's exam and was commissioned a 2nd Lieutenant in 1939.

Company M operated out of the armory located on Park Street in the center of Adams. In addition to its castle-like façade and intense focus on military readiness (training, drills, and maneuvers), the armory also provided a social center for the town's residents. Curly was involved in many of its activities, including military dances, raffles, and socials.

Curly was hired at General Electric in 1936 and commuted to

Pittsfield with his slightly used, gray four-door sedan on which he could be found tinkering most weekends. Curly would work as an assembler in the transformer department until he left for the war.

He was also a member of the Republican Town Committee. In 1939, with an interest in civic affairs, he decided to run as the Republican candidate for a one-year position on the Adams School Board.

During all his activities in town, between football, Company M, and civic interests, he met Veronica Susan Tarsa, a young woman who had moved to Adams as a teenager. She had attended North Adams schools, including Drury High School, a football rival of Adams High.

Curly was smitten with "Van," and the popular couple began dating. They became engaged, and eventually married at nine a.m. on July 4, 1940, at the St. Thomas Rectory in Adams. After a honeymoon trip to Canada, the couple resided with Curly's mother, "Ma," on Columbia Street.

Van continued working in administration at Arnold Print Works, and Curly at General Electric. A local newspaper describes the newlywed couple: "both are well known and popular . . . ." (*Transcript* 7-5-1940). Family, pleased with the union, described Van as sweet, loveable, and kind and Curly as gregarious, outgoing, and the kind of person you wanted to be around.

No one realized "the winds of war" were not far away.

Company M's troops trained that summer in Potsdam, New York, undergoing field maneuvers, drills, and firing exercises. The *Transcript* from August 19, 1940, names the many relatives and friends who visited the men at camp, including Van, Florence, and Richard Burnette, Curly's brother. It also notes the "people of Adams [who] donated cigars, cigarettes, and cookies to Company M."

The unit's training increased in 1941, with trips in January to Camp Edwards on Cape Cod and then to Fort Devens, Massachusetts. In March, Lieutenant Burnette left to attend a three-month US Army Officers School at Fort Benning in Georgia. He returned in June, rejoining the 104th Infantry at Camp Edwards, and they participated in more maneuvers in the Carolinas during the summer and fall, then returned north just before Pearl Harbor on December 7, 1941.

After spending months at Camp Edwards, in May 1942 Curly (now Captain Burnette) was reassigned from Company M of the 104th Infantry to the 317th Regiment of the newly reactivated 80th Infantry training new enlisted soldiers.

It's notable that an Army division consists of 14,000 men and comprises three regiments. Each regiment has approximately 3000 soldiers; the remainder are supporting arms, e.g., artillery, tanks, supply units, etc. Within each regiment are three battalions with approximately 1000 men per battalion. Over time, Curly would eventually lead the 1st Battalion of the 317th Regiment into battle.

Major Burnette, after training and before deployment to France. *[Courtesy of Tarsa/Burnette family.]*

Recruiting and training continued simultaneously for over a year, as the ranks were filled and civilians were turned into soldiers intended to defeat the German Wehrmacht. In April 1943, President Franklin Roosevelt visited and reviewed the troops of the 80th Division before they departed Camp Forrest to the Tennessee Maneuver Area for large-scale unit drills. Captain Burnette was promoted to Major in May.

To further harden the men in August, the division moved to Camp Phillips in Kansas for additional training, and the bleak landscape provided bitter cold weather, pot-bellied stoves, and frigid outdoor latrines. Major Burnette moved with the troops to the California-Arizona Maneuver Area in the Mojave Desert for 13 more weeks of training under another type of harsh conditioning, rough terrain.

By April 1944, it had been two years and four months since

America declared war. Eager to do his part, well-trained, and more than ready for combat, Major Sterling Burnette and the 80th "Blue Ridge" Division were moved to Camp Kilmer, New Jersey, to await deployment overseas.

The Allied invasion of Europe began on June 6, 1944, known as D-Day. Later the same month, the 80th Division boarded a troop train, which moved them to the Queen Mary for a voyage to Europe. There were stops in Scotland and England before the unit landed in France on August 1, 1944, almost two months after D-Day. The division was assigned to General Patton's 3rd Army.

When the 317th landed in France, Major Burnette was appointed Executive Officer of the 1st Battalion. Although a staff officer, he was seldom at his desk. Described as a thorough and efficient tactician, he was constantly in the field, encouraging junior officers and men to check on the progress of operations and combat conditions.

The 1st Battalion was in action almost immediately. Destroyed vehicles were scattered everywhere, artillery pieces and vehicles on their side, some beside the road, others blocking roads. Unburied German soldiers and farm animals lined the road. Bloated, bursting bodies and the scent of death permeated the air.

In mid-August, the aggressive regiment began taking casualties as it pushed the retreating and often counterattacking German forces across France through strange-sounding villages, towns, and cities like Avranches, Argentan, Évron, Sillé-le-Guillaume, and Châlons-sur-Marne. Other names were familiar from World War I battles, like Apremont and Saint-Mihiel.

Despite many contested river crossings, the 317th Regiment forged on, contending with sudden artillery, mortar barrages, and surprise assaults and ambushes by Panzer tanks, as well as bridges blown up by the retreating forces. Casualties mounted as their dogged pursuit continued. Ammunition, supplies, and replacements for the wounded and killed were a constant demand.

In late August or September, the regimental commander replaced his underperforming 1st Battalion commanding officer with its competent and well-respected executive officer, Major Sterling Burnette, who received a battlefield promotion to Lieutenant Colonel.

NOTE: During September 1944, the 317th Infantry Regiment suffered over 3000 casualties due to continuous fighting since August. The regiment essentially turned over its entire complement of men, in losses equivalent to 20% of the 80th Division's total casualties during the whole war (*One Hell of a War,* p. 93). The 1st Battalion was credited with capturing or killing hundreds of the enemy; for unswerving devotion to duty and mission as part of the 317th, it received a Presidential Unit Citation for Distinguished Unit.

Aside from the continued assault on the troop's senses, weather conditions worsened in autumn with the beginning of the rainy season. The soldiers were miserably cold and wet for extended periods of time. The weather turned roads into morasses, clogging vehicle passages, coating boots in mud, and making weapon maintenance difficult.

The most important item to an infantryman is his feet, which were now constantly wet from the rain, wading in rivers, walking in mud, or sleeping in water-filled foxholes. Trench foot became a serious enemy, just as dangerous as German gunfire. Good platoon sergeants ensured their troops changed socks daily to prevent medical evacuations.

The 1st Battalion had been proceeding east to west through the towns of Pont-à-Mousson, Vigny, and Luppy, heading toward the bridge at a crossing in Han-sur-Nied, a small village in France.

As of early November 1944, the 1st Battalion had been in continuous combat for over three months. The soldiers slogged through mud cautiously, trying to avoid the numerous anti-personnel mines that the retreating Germans had left on both sides of the road.

The Germans employed small, wooden anti-personnel mines, called Schützenminen, that contained a ½ pound explosive block. They were difficult to detect because they only used a small amount of metal. The mines were known as "shu" or shoe mines, approximately the size of a shoebox, and with just enough explosives to blow off a soldier's foot—however, in slippery conditions, men would fall on a mine, and it would destroy their upper torsos, killing them. The German Army, although retreating, used thousands of these mines to slow the advancing forces, and they extracted heavy casualties.

Often when under fire and in a minefield, Allied infantrymen followed the trail of their dead and wounded to get themselves safely through the minefield.

On the night of November 10, 1944, the regimental commander ordered the bridge over Han-sur-Nied to be captured, with Lt. Col. Burnette's battalion to lead the way. Everyone understood the bridge would provide a vital crossing to allow General Patton's armies to pursue the retreating enemy.

As Curly and his battalion followed up and began approaching the bridge, it came under heavy artillery fire from the hastily retreating 128th German Infantry regiment.

As a prelude to his attack, Lt. Col. Burnette had requested an artillery strike, which quickly scattered enemy vehicles and bridge guards who neglected to blow up the pre-wired bridge.

He immediately directed a tank platoon and a decimated group of Company A soldiers to assault the short, wooden, narrow bridge. In response, the American forces were quickly hit with automatic weapons, mortars, tanks, anti-tank guns, and most deadly—quadrupled barrel 20-mm German anti-aircraft guns, featuring high explosive rounds capable of causing numerous wounds with sniper-like accuracy.

The German crew-served 20-mm anti-aircraft guns fired high-explosive, highly accurate, deadly fragmentation shells. The guns were designed for large targets, such as buildings, vehicles, and anti-aircraft as protection for soldiers, but they could be very effective against infantrymen when lowered horizontally. The four guns, mounted on a large pedestal and manned by two soldiers, could spew out at least 400 rounds a minute over 2200 meters. The flak guns were much closer to Curly and his men as they tried to cross the bridge.

Records indicate that 1st Battalion Commander Lt. Col. Sterling Burnette, who led the advance on the Han-sur-Neid bridge, received mortal wounds to the chest and head from the 20-mm flak guns on November 11, 1944, which was, poignantly, World War I Armistice Day. The weapons firing from the commanding high ground on the east side of the river killed Burnette, as well as many infantrymen and tankers, and disabled several tanks.

Not to be dissuaded, the remaining infantrymen reached the other side of the bridge's span and, using wire cutters, severed the explosives' wires, and amid a barrage of shell fragments, the remaining 18 men took control. They were quickly joined by three tanks and Companies B and C, and held the bridge until reinforced by the 2nd and 3rd Battalions late in the day.

The regiment slogged on and continued to fight and win some of the war's fiercest battles until the war in Europe ended six months later—only without Curly.

Back home in Adams, the mail seemed to be slow. Van hadn't heard from Curly for a while; however, that wasn't too alarming, with his unit always moving from one town to another. Then on Wednesday, November 22, 1944, she received a Western Union telegram from the Acting Adjutant General stating, *"I regret to inform you your husband was seriously wounded in action in France eleven November until a new address is received for him . . . "* directing her to send his mail to a central postmaster address and she would be kept advised of his condition.

Sterling had been dead for eleven days.

Although they understand he is reported to be seriously wounded, Curly's family is anxious but heartened that he is alive. Eight days later, on November 30, 1944, another Western Union Telegram was delivered in Adams regarding Curly, this time to the wrong address, which delayed its arrival to Van. This one stated:

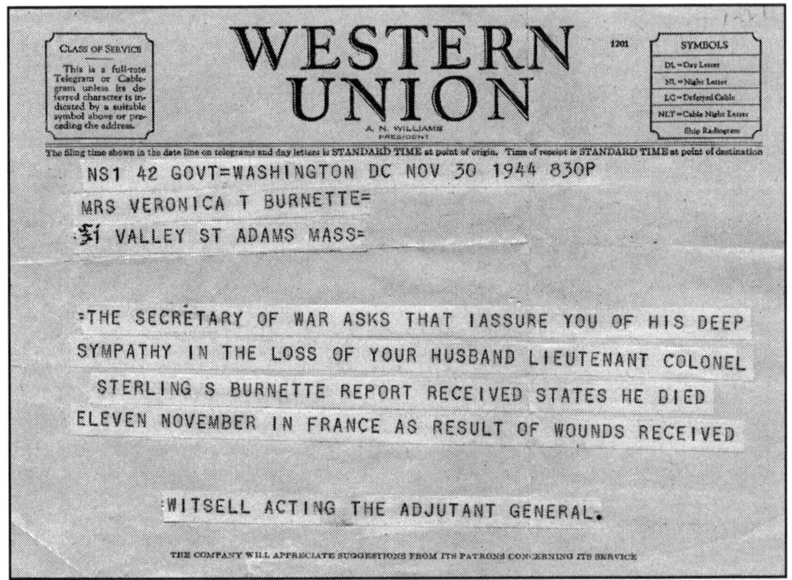
WESTERN UNION

NS1 42 GOVT=WASHINGTON DC NOV 30 1944 830P
MRS VERONICA T BURNETTE=
51 VALLEY ST ADAMS MASS=

=THE SECRETARY OF WAR ASKS THAT IASSURE YOU OF HIS DEEP SYMPATHY IN THE LOSS OF YOUR HUSBAND LIEUTENANT COLONEL STERLING S BURNETTE REPORT RECEIVED STATES HE DIED ELEVEN NOVEMBER IN FRANCE AS RESULT OF WOUNDS RECEIVED

=WITSELL ACTING THE ADJUTANT GENERAL.

[Courtesy of Tarsa/Burnette family]

The last news devastated the family, who thought Curly was recovering at a hospital in France. The word quickly spread, and the small town that had lost several of their sons now had yet another one, a very popular one, to grieve.

Curly's older sister Carolyn, who fainted when she heard the news of his death, had just delivered a son the day before his death. She was planning to name the baby Richard Sterling Cichetti, but reversed the names in memory of Curly.

Shortly before the war ended, the family was notified in March 1945 that Curly's heroic actions at the Han-sur-Nied bridge had earned him the Distinguished Service Cross, the nation's second-highest military honor. The citation was issued by the command of Lieutenant General George Patton on January 22, 1945.

It read:

Lieut. Col. Sterling S. Burnette 1st Battalion, 317th Infantry Regiment, 80th Infantry Division, United States Army. For extraordinary heroism in connection with military operations against an armed enemy. On November 11, 1944, the Battalion commanded by Col. Burnette was attacking against fierce resistance toward the town of Han-sur-Nied, France.

Col. Burnette was with the leading elements of one of the companies that had pushed forward of the Battalion to capture intact the one remaining bridge across the river in that sector. From his commanding position in front of the Battalion, inspiring his men by determination and bold, aggressive leadership, Col. Burnette fearlessly led the assault on the bridge through an intense concentration of enemy artillery, small arms, mortar, and machine gunfire. He fell mortally wounded at the approaches to the bridge, but his men, inspired by his heroic example, pressed on and secured the objective. Col. Burnette's conspicuous courage, heroic determination and supreme devotion to duty exemplify the highest traditions of the military forces of the United States.

Time passed slowly for the grieving family, and it was to be a year later (approximately late spring 1946) when they would receive Curly's personal effects. The War Department officially notified Van of Curly's temporary burial location in France. The correspondence stated that he was interred in a "US Cemetery, Limey, France, plot E, row 5, grave 103." Curly was located about 20 miles from where he fell in battle. The letter also stated, "It is anticipated, in the near future, the War Department will receive authority to return the remains of your husband, at Government expense, to the final resting place which you select."

In August 1945, Van had received a letter and a picture from a S/Sgt Smith showing her an area in Fussen, Germany, that the

Battalion had named after her husband. He signed the letter "One of Curly's Boys." S/Sgt Smith was awarded the Silver Star medal for bringing back the mortally wounded Burnette from the Han-sur-Nied bridge while under fire.

The family was notified in the summer of 1948 that Curly's body was to be returned to the United States for burial, and preparations begin. He was transported from France aboard the US *Oglethorpe Victory*, then carried aboard a train from Schenectady, New York, to Pittsfield, and arrived in Adams on July 30, 1948.

His body was met at the station by the men of Company M and escorted to the armory, where it was to lie in state until the funeral and then be buried in the family plot on nearby Maple Street.

The funeral became a personal affair for the town. Hundreds of citizens, former officers, and enlisted men of Company M stopped at the armory to pay their respects, in the same spot where he spent so much of his time as a National Guard soldier.

At the armory, a special testimony was given by "Dr. H. Fishkin, the medical officer who exerted all of his medical skill to save Curly's life after he fell mortally wounded with shots in the head and chest." Dr. Fishkin, who served with Lt. Col. Burnette both in the United States and overseas, said: "'Curly' was one of the finest officers I ever served with, and more than that, he was one of the best men I ever knew, and after all, what greater tribute can you pay a man than that. Had we had more officers like 'Curly' Burnette, there would have been far less men lost, and the War would have been over much sooner."

He went on to say, "Curly never should have been there exposing himself to danger as he did. That was not his job. He should have saved himself for the sake of his value as an officer, but that was Curly. He felt he must lead his men where they were to go, and if he were to speak to us now, he would tell us it was worth it" (*Transcript,* undated article titled "Funeral Ceremony").

After the testimony, Curly's body was escorted to the First Congregational Church, where the local pastor officiated at the last rites ceremony and read the 23rd Psalm; at the end of the service, Company M sergeants served as pallbearers, carrying his body outside.

Hundreds lined Park and Maple streets, paying final tribute as his body was escorted from the church to the nearby cemetery. At the cemetery, the Army Chaplain conducted graveside services. Then the honor guard and firing squad (comprised by officers who served with Curly) made their presentations, the bugler played "Taps," and Van was presented with the flag covering Curly's coffin.

In 1953, Veronica "Van" Burnette was officially presented with Curly's Distinguished Service Cross, in a special ceremony by Major General Allen, the Commanding General of Fort Devens. Curly's mother, brother, sister, and other relatives were present at the ceremony honoring him.

Lt. Col. Sterling Burnette and a fellow veteran, Private Walter J. Bednarz, were the only two Adams men awarded the Distinguished Service Cross. Both were killed in action during their heroic deeds. Adams honored them by having the Maple Grove Bridge and the Adams VFW Post named after them. Their pictures were also prominently displayed in the Adams Free Library.

Florence "Ma" Burnette, Curly's mother, became a member of the VFW Auxiliary and a Gold Star mother. She passed away in 1977 at the age of 93.

Veronica "Van" Burnette served as a longtime member and the president of the VFW Ladies' Auxiliary and was also a member of the American Legion Auxiliary. Van became an active member of the Grey Ladies, a volunteer non-medical group associated with the American Red Cross who visited the sick and injured during the war. She and Curly never had any children, and Van never remarried. She was heard often wistfully saying, "I couldn't think of anyone better than Curly." She passed away in 2008 at the age of 92.

Although it was recorded in the book *One Hell of a War*, few truly knew the horror of the World War II battlefield. After three months of continuous warfare, seeing his men killed daily, while living in the

In the Maple Street Cemetery, Curly and Van are reunited. *[Courtesy of the author]*

muck, mud, and desolation of war, the night before his death, Curly, after receiving his orders to capture the bridge at Han-sur-Nied, is recalled to have said to his friend, a fellow battalion commander,

> *"So long, Jim, I won't be seeing you again. I'll be killed tomorrow." He was normally cheerful, and a fun-loving man (said Jim). Therefore, I was instantly alerted when his attitude seemed so gloomy. I tried to cheer him up by noting all of us usually felt that we would get hit during an attack. He persisted and told me this was different—he knew he would die! No amount of talking would assuage his deep depression, and we parted with his reiteration of impending death. Sometime later, my XO told me that Curly had been seriously wounded. In fact, he had been mortally wounded. . . . "*

Colonel James Hayes, the book's author (2nd Battalion commander), later wrote: "I lost a treasured friend who died a hero's death."

Despite feeling strongly about this premonition of death, Lt. Col. Burnette rallied his men and led them valiantly to capture a critical bridge with complete disregard for his own life. Sadly, Van, Ma, and the Town of Adams lost a husband, son, and citizen. The nation, of course, gained a hero.

Sergeant First Class Daniel H. Petithory
Special Forces/Silver Star
AFGHANISTAN

Only weeks after 9/11, Dan's small special forces team was covertly inserted by helicopters under cover of darkness into southern Afghanistan. They were Company A, 3rd Battalion, 5th Special Forces Group (Airborne), Fort Campbell, Kentucky, charged with training, arming, and helping local mujahedeen led by Hamid Karzai, future president of Afghanistan. The mission was to overwhelm hardened al-Qaeda and Taliban forces that had harbored Osama Bin Laden, the perpetrator of the attack on America.

Anticipating weeks of preparation, the team saw their circumstances change quickly. The eleven-man team with a rag-tag group of mujahedeen suddenly had less than 48 hours to prepare for an onslaught of an estimated 1000 heavily armed Taliban, determined to sack the town and kill all of its occupants.

DAN WAS PREDESTINED TO become a warrior, but let's start at the beginning.

Dan's story begins years earlier in June 1969, when he was born to Louis and Barbara (LaRoche) Petithory in Pittsfield, Massachusetts, and then brought up in the small nearby town of Cheshire, a picturesque location nestled within the valley of the Hoosic River. The Appalachian Trail crosses through the center of the town.

With fewer than a thousand families occupying the town, it is no exaggeration to say that it was a close-knit community. Dan's parents had been long-time area residents. Louis graduated from

Pittsfield High School and attended Rensselaer Polytechnic Institute, then worked for many years at General Electric Company and *The Berkshire Eagle* newspaper. He was also a member of the local National Guard unit and had been a ranked chess player.

Barbara graduated from Adams Memorial High School, worked for a time at Wall Streeter Shoe Company, then graduated from McCann Technical School of Cosmetology and worked as a hairdresser for many years as their family grew to three children: Michael, the eldest, then Dan and Nicole.

Growing up, through sixth grade, Dan attended and walked to the nearby Cheshire Elementary School. During one of the parent-teacher conferences, Dan's mother distinctly remembers his teacher telling her that Dan was a good student but could be distracting to others since they sometimes listened to him more than to her. Reflecting on it later, she realized the conversation pointed to his budding leadership potential.

Dan was always on the move, keeping an eye out for feasible acts of daring. One wintry day, his friends were sliding on their stomachs down an icy path, and Dan said, "Let's stand up and slide." Not long after, he fell and jammed a lower tooth through his lip. He took the game of hide-and-seek seriously and was seldom found, often hiding in the 100-foot pine tree outside their house, or on a neighbor's roof.

In his youth, Dan was an altar boy at nearby St. Mary of the Assumption Church, where he served morning and Sunday Mass for Father Tom. (His parents had been married at Assumption.)

As a child, he played with Matchbox cars, then quickly grew enamored with GI Joe, and finally moved on to his dad's National Guard equipment stored in a duffle bag in the attic. Before long, he was playing soldier with the neighborhood kids, smartly outfitted with a canteen, entrenching tool, and web belt. When he got a little older, he organized paintball games in the woods near his home, always trying to improve his marksmanship.

Dan attended Adams Memorial Middle School for seventh and eighth grades and played Little League baseball and junior football.

As he entered Hoosac Valley High School in nearby Adams, his big interest was martial arts. In the upstairs bedroom he shared with his brother Michael, Bruce Lee's poster held a prominent place.

When Dan took an interest in something, he was all in. He began studying the Fred Villari style of Shaolin Kempo Karate, taking local classes. (The Villari technique combines Japanese and Chinese martial arts with western boxing.) It wasn't long before his family watched him compete and win his class in a tournament in Rhode Island. They once took a memorable trip to New Jersey to buy martial arts equipment that was difficult to purchase in Massachusetts. They got lost and ended up driving through Chinatown. The memory is now poignant; during their wanderings they recall looking in the distance and being awed by the sight of the twin towers, the loss of which not long after would precipitate Dan's deployment.

When they found the martial arts studio, they bought several nunchucks, a three-sectional staff, a twin Japanese sais, and some shurikens (sharp throwing stars) before returning home. Dan was delighted with the purchases; Barbara less so, particularly when he began using their lone pine tree for a shuriken target.

Dan was never still very long. He played goalie on the high school's intramural soccer team, eventually injuring his ACL, which would later plague him on special forces missions. He worked at Joanie's Ice Cream/Grill on Route 8 and, for a time, had a paper route. He and Michael shared their parents' hand-me-down vehicles and could often be seen in a white 1979 VW diesel Rabbit.

Dan loved music and, on weekends, disc-jockeyed dances at Turn Hall in Adams, deftly handling graphic equalizers, mixing boards, amplifiers, and lighting equipment. This talent would serve him well in his future communications job with the special forces.

While in high school, he would visit his older brother Michael at UMass. On one visit, he helped engage hundreds of students in a vast snowball attack on nearby Amherst College that turned into a melee. Sixteen-year-old Dan was up front, leading the group with "Follow me!"

Often after a big hometown basketball game between rival cities Adams and North Adams, Dan and friends would balance a deflated basketball on top of the twenty-foot statue of President McKinley in downtown Adams. (The first year after Dan's death, his friends reinstated the basketball prank. The local police understood it was a tribute to Dan and said they would leave it there for a while.)

It seems like Dan was the daring one in the family, whether taking a canoe too far out on the lake during a family vacation and having to be coaxed back to shore, or accidentally shooting a hole in the family's floor with his grandfather's shotgun. Luckily, for many years only his brother knew of the spot; he'd been able to coax the carpet fiber back together so Mom wouldn't find it.

There was never any doubt where Dan was headed when he graduated from Hoosac Valley High School in June 1987. By September, he was in the US Army, having completed basic training, then attending a seventeen-week Law Enforcement Training Course at Fort Leonard Wood in Missouri, after which he became a military police specialist.

Dan was transferred to Fort McClellan, Alabama, where he was promoted and became part of the base's Special Reaction Team. As he looked for more challenges, he was able to volunteer and attend the challenging Air Assault School at Fort Rucker, Alabama.

After several years of assignments, Dan's unit was thrust into action in 1990 when Iraq invaded Kuwait. His military unit spent weeks preparing and then some time in-country supporting Operation Desert Storm in the liberation of Kuwait, joining a thirty-nine-country coalition of allied forces. His detachment worked with thousands of detainees, conducted local patrols, and aided with convoy security.

The forty-three-day battle in early 1991 quickly became a rout, with intense ground action lasting only 100 hours before Iraqi troops were decisively pushed back into their country.

Upon Dan's return to the United States, he completed a three-week Basic Airborne course. Expressing a continued desire to be with the warriors after seeing special forces in action during Desert Storm, Dan applied for and was accepted into the Army's special forces rigorous assessment and selection phase to determine if he would qualify for the intense Special Forces Q (Qualification) Course.

Dan successfully passed all phases and was assigned to Company A, 3rd Battalion of the 5th Special Forces Group at Fort Campbell, Kentucky. This group's Area of Responsibility (AOR) was the Middle East, a broad area including Central Asia, the Persian Gulf, and the Horn of Africa.

Dan, enthused about being assigned to A-Teams. *[Courtesy of Petithory family]*

It is useful to note that there are seven Special Forces Groups, each responsible for different areas of the world. Each group has three battalions, and each battalion has three companies. Each company has six Operational Detachment Alpha (ODAs—called A-Teams).

Dan was assigned to Operational Detachment (ODA) 572. Twelve team members were led by a captain, a warrant officer, and a team sergeant, with nine individuals specializing in weaponry, demolition, communications, medicine, and intelligence. Most would be cross-trained. He had been given the military occupational specialty (MOS) of communications—known by code 18E—and the rank of sergeant.

At ODA 572, training was demanding, and Dan knew his performance would be under scrutiny for months.

In 1992, as a communications sergeant, Dan spent almost the entire year training. For twenty-four weeks, he attended a communications course. The training was difficult, studying complex global communication systems. At one point, he mentioned in a letter to his mom that he was having difficulties. Yet by the time the twenty-four-week course was over, he had a solid understanding of radio and satellite systems and how to communicate with many different support groups, whether they were mortar teams, jet fighters overhead, or B-52 bombers miles away. Dan became a master at his job.

With a Middle East team assignment for the remainder of the year, he spent twenty-two more weeks learning Farsi, a language commonly used in Afghanistan and Iraq. Months later, the team sergeant would send six Farsi-speaking team members to Limerick, Maine, for immersion studies. They would rent an out-of-the-way Airbnb, hire three Farsi instructors, and hold immersion classes for several weeks.

During the 1990s, Dan's A-Team was continually supporting democratically inclined governments throughout their Area of Responsibility. They were often called to train government soldiers in small-unit tactics and weaponry to strengthen their government's response to local insurgencies.

The A-Teams usually rotated in and out of the country for periods of sixty to ninety days, often repeatedly. Since most of the work was confidential, Dan's family seldom knew where he was, which always created a slight anxiety factor, but they were delighted to see him every year or so.

During one of his visits, he left home with his backpack filled with rocks to go hiking. His mom noted that it was raining out. Dan's reply was, "If it ain't raining, you ain't training!" It was one of his favorite sayings. The communications gear he carried on his back weighed well over 120 pounds, hence the simulated pack of rocks.

As he learned his job, the 6 foot 4 inches, 225-pound Green Beret established a reputation as one of the company's best "commo guys." Dan's discipline, mental toughness, native intelligence, fitness, and flexibility made him an excellent fit for the team. There were times under severe overnight training conditions when he would stay up all night to ensure his team's communications were locked in.

Because of his size, his teammates joked that Dan may not be the fastest on their Friday ten-mile runs, but he sure could carry a lot of gear.

His sense of humor and good nature attracted others to him. He was also known as a morale builder and the life of the party. Many called him Dan-O. One member recalls that on first meeting, Dan introduced himself as "I'm Daniel H. Petithory, the H stands for Hero" and they both had a good laugh and became close friends. The same team member also said, "Often when we had some downtime, [he] would take charge of our entertainment, and he always had us watch the DVD *Reservoir Dogs*, one of his favorite movies about a jewelry robbery."

Some months after Operation Desert Storm, ODA 572 returned to Kuwait several times, training the Kuwaiti military in self-defense tactics. Dan, constantly experimenting with his comms, noticed a boy with a kite and asked him to hook an extended antenna to it. With a limited "line of sight" radio, he found he could reach over a hundred miles back to Kuwait City.

ODA 572 was also involved in Operation Uphold Democracy in Haiti in the mid-1990s, when the United States sent tens of thousands of troops to establish order and enforce the results of a democratic election. Dan and his team spent about seven months in Haiti monitoring the country's first free election and helping to suppress violence.

While Dan and some of his team were deployed near Cap-Haïtien monitoring the election, he had the idea of trying to create a memorable moment for the group. Taking a vintage Vietnam-era radio, he laid its extended antenna in the ocean and was able to contact a surprised HAM operator in Montana, allowing each team member to make a short "morale" call home.

The team was deployed several times to Pakistan to train its Special Services Groups in small arms, unit tactics, and airborne drills. One bitter wintry night, Dan stayed up to ensure his group had comms with the outside world. Even in this arduous case, he defused the misery with humor, reporting, "This is the Donner Party checking in . . . ."

Dan, deployed with his team to Pakistan, circa early 1990s. *[Courtesy of Petithory family]*

ODA 572 also deployed to Kenya repeatedly, training the 20th Para Battalion in small unit tactics and airborne operations. Dan was always trying to lighten things up. One day, he laid some detonation cord to scare away huge flocks of annoying, scavenging crows that had been constantly pestering the group. The "det-cord" trap only resulted in one bedraggled casualty, and the birds were back in force the next day.

The group also spent several extended periods training the military of Kazakhstan. After exhausting long days spent out in the field, on weekends they would relax at a small shanty of a bar in the nearby town and have a few beers. The place was moribund, featuring poor music and a slow disc jockey. Finally, one Friday night, Dan arrived with a sleeve of his CDs. He nudged the DJ aside, started playing his own repertoire, and got the whole place dancing and jumping up and down. After that, on Friday, everyone started looking for "DJ Dan," and someone listening in might hear something like this: "Rocking Dan Petithory coming at you from the number one radio station in Kazakhstan . . . and we take requests all day long!"

Pictures show the 6 foot 4 inch, 225-pound Green Beret with a huge smile, wearing a distinctive Russian fur hat with ear muffs, dwarfing the local crowd, certainly winning the hearts and minds of the local people.

In the late 1990s, Dan received his HALO certification and moved to ODA 574, where he would remain in the Special Forces for the rest of his time. HALO is a parachutist designation meaning High Altitude Low Opening. It refers to parachuting from a plane at an altitude above 20,000 feet and making a rapid descent before opening the chute at around 3,000 feet to avoid enemy radar detection or ground observers. It is a dangerous and physically demanding

Dan, in center, with Kazakh soldier on maneuvers. *[Courtesy of Petithory family]*

Creating alliances and supporting fledgling democracies. *[Courtesy of Petithory family]*

craft and is usually performed in the most select of circumstances. Dan, being Dan, even in these perilous exercises, would bring a little levity into the plane, sporting Elvis sunglasses.

Dan was in Kazakhstan on September 11, 2001; he and his team had been tasked again with training the country's soldiers in counter-insurgency tactics. They were living in dilapidated Russian barracks when they got word of the attack on the twin towers and the Pentagon. Wanting to act, to respond to the perpetrators immediately, they impatiently waited for orders. Reluctantly, they returned to the States for direction.

The President declared the "War on Terror," and the Joint Chiefs of Staff began planning for unconventional warfare, a war fitting for the Army's Special Forces. In the 5th Group's AOR, planning began immediately. The broad objective was to destroy Afghanistan's Taliban and al-Qaeda safe havens, where the planning and support for 9/11 originated. The A-Teams were directed to bring the fight to the enemy.

So began Operation Enduring Freedom. In October, Dan and his team (ODA 574, also now designated TEXAS 12) entered an isolation facility to prepare for war. In preparation for missions, A-Teams are often sequestered from the outside world where they create an in-depth action plan, privately reviewing maps, intelligence, photographs of warlords, enemy locations, numbers, weapons, and structure.

Once preparations were complete, the team left the US and was flown to a secret base in Uzbekistan known as K2, or Camp Stronghold Freedom. Shortly afterward, they were secretly inserted by Apache helicopters under cover of darkness into Afghanistan. The United States had dispatched 9 or 10 teams throughout Afghanistan. It was mid-November 2001, and they became the first team in southern Afghanistan.

After connecting with a local leader, Hamid Karzai (who would, in time, become the President of Afghanistan), their goal was to help him train, arm, and unite local tribes. Alongside his *mujahedeen*, they would reclaim the city of Kandahar from the Taliban. CIA operatives also accompanied the team.

The men were heavily laden with weapons and equipment. Each carried an M4 assault rifle, M9 pistol, grenades, ammunition, personal items, and the particular gear associated with his responsibilities, ranging from communications equipment to medical supplies to specialized weapons. Individual packs easily weighed over 100 pounds; Dan's far exceeded that with the communications equipment.

Initially, the team's mission was to establish contact and protect Karzai, who was being hotly pursued by the Taliban. They would also equip and train Karzai's supporters. Then, after some months, their aim was to occupy the small town of Tarin Kowt, eventually gaining additional territory with the hopes of capturing Kandahar, about 70 miles away, and toppling the Taliban regime.

Unexpectedly, within hours after the team arrived, they received word that the citizens of Tarin Kowt had risen up and killed the Taliban mayor. The team's timetable was suddenly jettisoned. Instead of having months to cultivate a fighting force, they had mere days to have weapons and ammo airdropped, train Karzai's motley militia, and get to Tarin Kowt to prepare for the anticipated vicious onslaught by the Taliban.

TEXAS 12, which had been inserted nearby, and Karzai's men headed together for Tarin Kowt in any vehicles they could commandeer. The Taliban had been searching for Karzai. Now they had a chance to capture and kill a potential competitor, punish a town for its audacity, *and* repel foreign invaders. To the Taliban, it was a grand temptation. Karzai's informants reported that the Taliban and al-Qaeda were sending a thousand men from Kandahar to retake the town and kill its inhabitants.

The situation quickly became a perfect example of the Green Berets' motto, *De Oppresso Liber*, To Free the Oppressed. The rumors of the fighting force caused anxiety for the small band of local *mujahedeen*, but the Berets tucked in and made preparations. The team set themselves up on a hill outside of Tarin Kowt with a long view of a valley approach, allowing them time to track and destroy approaching enemy vehicles. It also gave them some space to retreat to the town if needed as their backup plan. Calling planes for overhead reconnaissance, they learned when vehicle movement out of Kandahar began toward the town, which happened almost immediately.

It wasn't long before Navy planes spotted eighty vehicles headed their way with an estimated 500+ Taliban on board. Dan became deeply involved with his laptop and radio, communicating with the overhead planes. The team signaled the pilots to begin their bombing as the vehicles quickly approached. The caravan was a jumble of Toyota trucks and old Russian troop carriers, even one with an artillery piece. The planes engaged them in some narrow passes leading to Tarin Kowt, hitting lead vehicles; the bombs flipped some trucks completely in the air.

A few trucks got through the cascade of bombs, and those were stopped by defensive positions the team had set up, manned by themselves and the *mujahedeen*. Dan was right in the thick of the organized chaos. Known for his coolness and knack for orchestrating air strikes and other supporting fire, he continued to call in targets.

"He loved adjusting mortars, artillery, and close air support. He loved his job. He loved his team. He loved all of it. Dan-o, as they called him, was still cracking jokes as ordnance was exploding all around him" (Moore, *The Hunt for Bin Laden*). Dan used a laser finder to calculate the distance to the target and then coordinated that with a handheld GPS system and satellite antennas, feeding the data to bombers overhead, all while under fire from the enemy. With satellite-directed bombs, the team of eleven and the *mujahedeen* destroyed at least 35 vehicles and killed 300 Taliban. The team, though outnumbered 50 to 1, had made a huge impact.

The Battle of Tarin Kowt was later considered the turning point of the war in southern Afghanistan. Many team members were awarded bronze stars and purple hearts for their wounds after the seven-hour fight.

Next, TEXAS 12 turned toward Kandahar, fueled by their success and increasing support from the local population. The Taliban remained in large numbers. Still, they were rapidly retreating, and the team continued to use air support for enemy reconnaissance.

On December 3, TEXAS 12 was just north of Kandahar when Dan again made use of his air support communications to overwhelm a larger force of Taliban. TEXAS 12 met fierce resistance on the outskirts of Kandahar, overlooking a bridge. The team had a number of close-running battles with numerous insurgents. Luckily, they continued to be supported by airplanes and prevailed in these gunfights. That night, exhausted, they set up security and fell fast asleep on the dirt in an area they had facetiously dubbed the "Alamo." It is notable that one team member had been shot in the throat but survived, and Dan, the best marksman on the team, volunteered to take his place the next day.

On December 5, 2001, TEXAS 12 connected with another Green Beret team. This Command team took over. As one of their first acts, they directed a penetrating bomb to destroy the cave where they suspected insurgents were hiding.

What happens next has surely been studied and reviewed time and again, from the most classified levels of the Pentagon to many warm, quiet households in the Berkshires. Using their own TACP (Tactical Control Party) specialist to call in bombing coordinates, the Command team directed a B-52 to drop a 2000-pound bomb. The GB-31 high explosive bomb requires a safe space of at least 500 yards from friendly troops. What the Command specialist didn't realize is that he accidentally gave coordinates within a hundred yards of the Berets' position.

It was a horrible mishap. The massive ordnance killed three Berets, including Dan, and wounded nineteen others. It also killed and wounded many Afghan allies and superficially wounded Hamid Karzai. Dan and his two fellow team members earned the tragic distinction of being the first special forces soldiers killed in the War on Terror.

Dan's hometown paper, the *Transcript*, reported on December 14, 2001, that he had been "one of the first four casualties in Operation Enduring Freedom." Just several days later, Kandahar fell to the allied forces.

On December 11, 2001, Louis and his son Michael Petithory walked across the tarmac at Albany (NY) International Airport to greet the airplane carrying the body of their son and brother. Two Green Berets accompanied them while seven other Berets carried the flag-draped white casket to the black hearse. Dan was brought home to Cheshire, Massachusetts, to the Dery funeral home. Calling hours for the family seemed to go on forever. So many wanted to pay their last respects.

Dan's funeral was held at a packed St. Mary of Assumption Church on December 13, 2001. Much had happened in just a week. Few will forget the brilliant piper from the Berkshire Highlander Pipe Band who played "Danny Boy" as the family entered the church. Several priests officiated, including Father Tom, whom Dan used to serve as an altar boy. Many dignitaries were there to honor Dan, including the state's governor, its senators, several Army generals, and a sea of his fellow Green Berets.

Senator John Kerry spoke at the service, as did Dan's team leader. Dan's sister, to whom he was very close, recalls how the church was decked out so festively for Christmas with wreaths and Holy Family statues in the background. It was resplendent yet incongruous.

She also remembers that when they left the church to go to the cemetery, elementary school children lined the road waving flags, a beautiful send-off for a local hero.

At the cemetery, Barbara was presented with the American flag that covered the casket by Major General Geoffrey Lambert, Commander of US Army Special Forces Command. A 21-gun salute was rendered, and "Taps" was played. A bagpiper again played "Danny Boy," this time from afar, on the hillside.

As a final gesture to Dan after most people had left the cemetery, his dad, brother, and A-Team members opened a bottle of Wild Turkey, passed around shot glasses, and toasted their dear friend and comrade. They left a glass and the remainder of the bottle with Dan.

After Dan's funeral, the Army awarded him the Legion of Merit for "exceptionally meritorious conduct in the performance of outstanding service."

Subsequently, a scholarship fund was created at Hoosac Valley High School in remembrance of Dan. A portion of the school's courtyard with a memorial garden designed by the students is dedicated in his honor. Not far from Dan's house, 3.5 miles of the Ashuwillticook Rail Trail is officially named the "Daniel H. Petithory Memorial Trail," in an area where Dan used to play. It feels as if his spirit still imbues the space.

In 2002, Colonel John Fenzel, Dan's A-Team Company Commander, addressed a military academy. He chose as his theme *Heroes*, and chose as his subject Dan. This is a portion of his address, and in truth, it says all that needs to be said:

> *I'd like to talk to you tonight about heroes.*
>
> *In working with our allies, [Dan Petithory] volunteered to go with me to Pakistan. . . . he parachuted into the Indus River valley out of a C-130 cargo plane right behind me, in the pitch darkness, with a 120-pound rucksack filled with satellite radios and ammunition. And through his extraordinary communication skills, he kept 200 American and Pakistani soldiers organized as we traversed the rough terrain. With six miles left to go, I can remember a distinct cracking sound behind me. Under all the weight he was carrying, the ground had given way, and he had fallen down a hidden ravine. He turned his ankle and tore some ligaments. His back was badly wrenched. He was in terrible pain. I told him that we would wait until some medical help arrived to evacuate him,*

but he refused and, despite his injuries, insisted we continue to move on, saying, 'Sir, you gotta play hurt.'

We shifted his load . . . kept moving . . . he never complained.

It strikes me Heroes are selfless, and hold themselves and those around them to a higher standard. On September 11, Dan volunteered to deploy to Afghanistan immediately . . . he knew the terrain, the people, their language. Dan helped organize Karzai's band of soldiers, helped to train them, and he went to war with them. He fought beside them and befriended them. He was truly a soldier–diplomat–ambassador for everything you and I stand for.

Just before Christmas (2001), Dan and two other Special Forces soldiers were killed in Afghanistan doing what they loved. For this country they loved so dearly, the result of a rare and steady dedication of one's lifetime to something larger than themselves. They knew the risks they faced. And all of us are more secure today because of their willingness to sacrifice themselves for a greater purpose so that we might remain free. . . . They give us direction and provide us with hope and inspire us to press on. And Dan Petithory . . . Dan Petithory is a hero of mine. Because of all that . . . and because he went first.

North Adams, Massachusetts

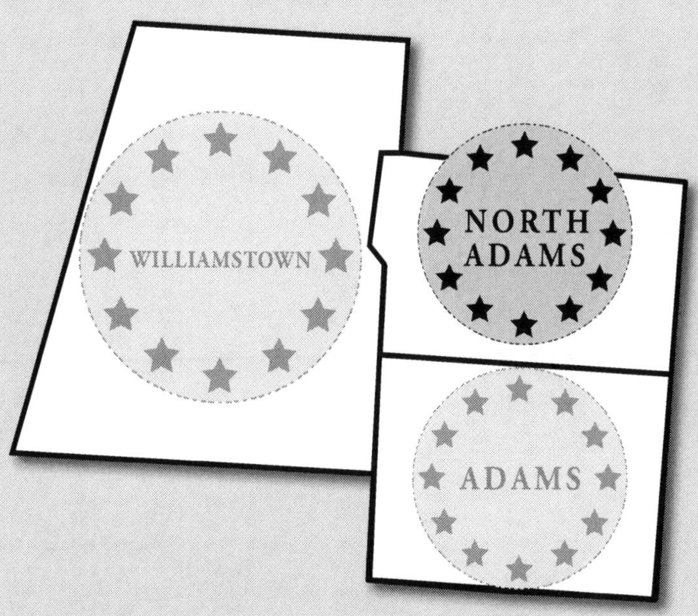

Elisha Nims
English Militiaman
French and Indian War

It was a hard blow to the lower part of his back from a .69-caliber lead musket ball, one-fourteenth of a pound, traveling at 800 feet per second. He was, in all likelihood, unconscious and paralyzed immediately from the severe mortal shot. He might have seen a wisp of smoke or felt a tug of his head after collapsing.

This is the story of Elisha Nims, a militia soldier who died in an ambush outside Fort Massachusetts in 1745 during King George's War (1744–48) and was formally buried 255 years later in the Soldiers' Circle of a North Adams cemetery.

Early History

THE FRENCH AND INDIAN War was a series of conflicts involving England and France. The two countries fought each other repeatedly for nearly a hundred years, meeting in four wars beginning in the late seventeenth century: King William's War (1688–1697), Queen Anne's War (1702–1713), King George's War (1744–1748), and the French and Indian War (1754–1763).

While many of the battles took place in Europe, the wars also involved the Americas, where each country was seeking to gain territory and the wealth of trade. The wars in the Americas, although part of a larger conflict, might have been better labeled the "English and French Wars."

Almost from the time of the first settlements in the new country, disagreements between France and England continued, primarily over territorial possession. As hostilities increased, both countries used Indigenous peoples as proxies to help them fight their battles.

In the mid 1700s, the number of English colonists had grown to almost 1,000,000, and Indigenous tribes had become disenchanted with the settlers who had laid to claim their lands, passed on devastating diseases, and made every effort to subjugate or eliminate them.

The French, with approximately 60,000 citizens in the Americas, had broader and closer alliances with the Indigenous Americans and controlled larger swaths of territory. They were more accepting of local cultures; learned native languages; and their missionaries, traders, and trappers lived and married within the tribes and were accepting of their customs. The French traders also gave their Native allies ammunition and guns (which the British refused to do) and were more adaptive to their ways.

The English settlers were supported by the Iroquois League, a confederation of six Native nations: Mohawk, Oneida, Onondaga, Cayuga, Seneca, and Tuscarora tribes. The allegiance was driven more by the league's distaste for the French than endearment with the British. The Delaware, Shawnee, Abenaki, and Mingo tribes, among others, aligned with the French.

Hostilities, surprise attacks, and ambushes became a way of life on the frontier, with English colonists defending themselves against predatory attacks sponsored by France from their New World headquarters in Québec.

In the 1740s, in response to France's efforts to win control of the Americas and also to limit incursions from the Dutch located in nearby New York, a series of English forts and blockhouses was constructed from Connecticut across western Massachusetts, described as "a string of wilderness defenses built to guard the Bay Colony from French and Indian hostilities."

Fort Massachusetts

In 1744, Massachusetts Governor William Shirley authorized the building of Fort Massachusetts in East Hoosuck, now known as North Adams, and a blockhouse in West Hoosuck, now Williamstown. The fort was constructed in a meadow adjacent to the Hoosic River. Its location was chosen because it intersected several major trails and fords traversed by hostiles. British soldiers and militiamen led by Captain Ephraim Williams, Jr., built the fort.

This boulder marks the former site of Fort Massachusetts. Note the British and American flags. *[Mike Remillard Photos]*

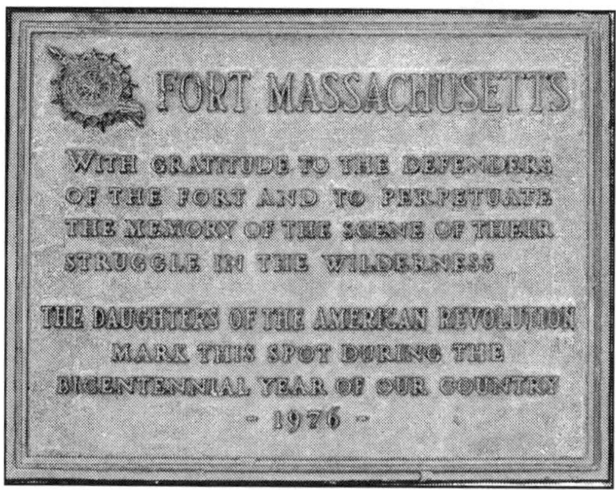

Detail of plaque. *[Mike Remillard Photos]*

Most forts were located miles from each other and patrolled by colonial militias, often accompanied by dogs to sniff out the enemy. The intervening areas were desolate and fraught with danger, patrolled by small groups of scouts and soldiers minimally equipped with rifles, scalping-knives, hatchets, and powder and shot. The clean-shaven militiamen wore loose linen shirts, leggings, sometimes leather jerkins, and traveled light, carrying only bedrolls and utensils.

It is said that the responsibilities of this wide-ranging group of soldiers were so dangerous and the predatory acts of hostiles so frequent that in the early 1700s, Governor Shirley's Massachusetts Colony authorized a horrific bounty for scalps of Native peoples: 40 pounds awarded for each scalp from males over the age of twelve, and 20 pounds awarded for those of women and children.

Fort Massachusetts provided a safe haven for early settlers from East and West Hoosuck. It was estimated to occupy an area of 80 feet by 120 feet with its logs stacked horizontally like bricks to a height of twelve feet. The logs harvested from nearby Greylock Mountain were trimmed to approximately 6 inches by 14 inches and of varying lengths with the first layer placed horizontally on a stone foundation, then stacked. The corners were dove-tailed and connected by wooden spikes.

Often forts had two watch-boxes constructed in opposite corners where sentinels stood watch. There is some evidence that Fort Massachusetts had one watch box and, opposite it, a two-story great house that served the same purpose with its bottom floor possibly serving as officer quarters or storage. A series of smaller barracks-like structures was built inside the compound to house soldiers, settlers, and their families as well as to store ammunition and powder. Construction also included the digging of a well for use by the fort's inhabitants.

Elisha's Ancestors

Elisha's grandfather, Godfray Nymn, born in 1648, emigrated as an indentured servant from Middlesex, England, when he was sixteen years old. Presumably arriving in Boston, he made his way to Northampton, Massachusetts, and records show that his name was anglicized to Godfrey Nims. As a youth, he had some trouble in Northampton and quickly moved to Deerfield where he settled.

Over time, Godfrey was married, widowed, and remarried, having eleven children in total. As a leading member of the community, Godfrey was part of the local militia and fought in the "Falls Fight" at what is now Turners Falls, Massachusetts, during King Philip's War.

Disaster struck his community of Deerfield just before dawn in February 1704, during Queen Anne's War, when the inhabitants were attacked and its population taken hostage or killed. During the massacre, seven of Godfrey's children were killed, deemed too young to make the rigorous 300-mile march to Montreal. Some were smothered, others dispatched with a blow to the head. His wife, Mehitable, weakened during the trek and was slain.

Ebenezer Nims, Elisha's father, was one of the four children to survive. Two other siblings were ransomed, and his sister never returned home, marrying into the Huron tribe. Unable to cope with the rigors of the massacre, Godfrey died shortly afterwards and was buried in the Old Deerfield cemetery.

While captive in Canada, Ebenezer (1687–1766) met and married fellow captive Sarah Hoyt (1686–1761), and they had a son before their freedom was purchased. They returned to Deerfield in 1714. Upon their return, they had four more children, one of whom was Elisha. He was born on June 20, 1720, in Wapping, a hamlet within the town of Deerfield, Massachusetts. It is interesting to note that the location of Fort Massachusetts is only 40 miles northwest of Deerfield.

Elisha's Life

Little is known of Elisha's early life, although it is presumed that much of it was involved with subsistence activities, like assisting his father in clearing land, building cabins, growing food, and hunting and fishing in the nearby forest and streams.

It seems that Elisha joined the militia in his early twenties; he first appears on the Fort Shirley (Heath, Massachusetts) muster roll, where he was involved in its construction and the patrolling of nearby woods. Elisha was listed on the roll as late as June 9, 1745, just days before the ambush outside of Fort Massachusetts.

At some point in early June, he was directed to proceed to Fort Massachusetts and covered the intervening twenty-plus miles in time to be involved in a skirmish with Native Americans on June 11, 1745.

On that day, Elisha was part of a detail of militia/soldiers guarding a work party, gathering water from a nearby spring near the north bank of the Hoosic River, about 300 yards from the fort. The hostiles, hidden from sight, surprised the group and succeeded in mortally wounding Elisha, seriously wounding another soldier, and capturing a third. Some records indicate that Elisha was scalped before a rescue party was able to drive off the ambushers, killing one of them.

Elisha, hit in the lower spine, was carried or dragged back to the fort. With no medical intervention at the isolated fort, he lingered only briefly, succumbing to his wound the next day. At the age of twenty-seven, he became the fort's first fatality.

It was later determined that his backbone had a .69 caliber musket ball lodged in it. The projectile likely came from a Charleville 1728 flintlock musket (.69 caliber) used by French forces during the eighteenth and nineteenth centuries and often supplied to their tribal allies during the French and Indian War. The muskets had a limited range of about 80 yards or less, indicating that the ambushers were hidden close by when Elisha was shot. (France would later supply colonists with tens of thousands of Charleville muskets during the Revolutionary War.)

There are records indicating that Sergeant Hawkes, the fort's non-commissioned officer, and probably others, used a Queen Anne .71 caliber flintlock musket imported from England. The flintlock rifle and often accompanying Queen Anne pistol were manufactured during Anne's reign, 1707 to 1714.

The soldier wounded and captured when Elisha was killed, Benjamin Taintor, survived fourteen months of captivity before being repatriated under a flag of truce back to Boston in 1747. During his captivity, he was tortured and at one point required to run through a long gauntlet manned by a hundred of his capturers, absorbing countless blows from fists, sticks, and war clubs.

After his death, Elisha was buried in what was called God's Acre, a small cemetery west of the fort. The ground contained soldiers and family members who died from disease or wounds while serving at Fort Massachusetts. It was the first cemetery in western Massachusetts, and its exact location has been lost to time. One record indicates that 14 soldiers and one "Stockbridge Indian" were buried at the cemetery.

The Siege

Fourteen months after Elisha's death, Fort Massachusetts was surrounded by over 800 French soldiers and natives from the St. Francois tribe of Canada. The besiegers were led by French General Pierre de Rigaud (the Marquis de Vaudreuil); a small garrison led by Sergeant Hawkes, who had been previously wounded, defended the fort. Hawkes and his group of twenty-two soldiers and five women and children initially resisted the besiegers, but dysentery and other illnesses, combined with a low supply of shot and powder, spelled the fort's doom.

Surrounded, under incessant gunfire both day and night, and subject to sniping from a nearby ledge that allowed the hostiles to fire directly into the fort, the British militia group reached out to talk with Rigaud. Hawkes agreed to terms with the French general and surrendered on August 20, 1746. During the battle, records indicate that one militia man was killed; estimates were that forty-five French and Native Americans died.

After discussion with the French commander, it was agreed that all captives would be prisoners of the French, children would stay with their parents, and prisoners would be exchanged as soon as possible. The weak would be carried and everyone could keep their clothing. At one point, under considerable protest, some captives were given to the Native Americans.

The fort was immediately ransacked and burned to the ground. The captives then marched to Canada, the arduous trek taking twenty-one days with half the prisoners succumbing along the route and others during captivity. The remaining fourteen were eventually ransomed and returned to the Americas.

Fort Massachusetts was rebuilt over the years 1746 and 1747 and again became subject to siege, this time by 300 Native Americans and 30 Frenchmen in 1748. During this siege, the defenders prevailed and the attackers retreated.

The original fort was built under the supervision of Ephraim Williams Jr., a captain in the militia. He was responsible for building and defending the series of forts and blockhouses stretching from Connecticut to western Massachusetts. Ephraim was absent when the fort was destroyed in 1746.

Interestingly, in 1747, Elisha's father, Ebenezer Nims of Deerfield, petitioned state authorities to reimburse him for his son's rifle that was taken by the enemy during the attack on Fort Massachusetts. Captain Ephraim Williams Jr. certified the request. The rifle was worth 20 pounds. The records do not indicate whether or not he was reimbursed.

Fort Massachusetts was rebuilt in 1747, used for some years, and in 1759 decommissioned following the Battle of Québec. Over time, with its structure falling into disrepair, the fort attracted nearby settlers who scavenged its timbers for their own uses and farmed the land on which the fort had stood, erasing most of its presence.

In the 1850s, a Williams College professor, excavating the site with a group of students, found the rough headstone of Elisha, along with his skeleton and a number of artifacts that were brought back to Williams College for study. Old records contain macabre information, notations that the skeleton had "large thigh bones," indicating Elisha may have been over six feet tall, and that the "skull was in perfect shape with only one tooth missing." At the time, a commemorative elm tree was planted in the middle of the former parade-ground of the original fort.

In 1895, the Fort Massachusetts Historical Society purchased the site. Williams College returned Elisha's remains and the headstone to the society in the 1940s; ultimately, the North Adams Public Library took possession of them in the 1950s. Over the years, the headstone disappeared, and Elisha Nims' remains are presumed to be reburied, with the exception of his backbone.

An historic postcard documents Elisha Nims' final burial site. *[Courtesy of North Adams Historical Society]*

In 1933, the fort was reconstructed by the Works Progress Administration, an agency that employed millions during the Great Depression. The construction featured a fieldstone cairn with Nims' tombstone. The fort was a tourist site during the 1960s, but due to

This historic postcard from 1950-1960 shows the Fort as a tourist attraction. *[Courtesy of North Adams Historical Society]*

lack of funds and vandalism, it was torn down in 1973. The elm tree died and the cairn disappeared. The reconstructed chimney and a plaque are all that remain.

Elisha's remaining vertebrae were passed between the library and the historical society. For a period in the 1990s, they were put on display at Massachusetts College of Liberal Arts and the North Adams Public Library, along with other artifacts of Fort Massachusetts including cannon balls, a defender's jackknife, and hand-forged nails.

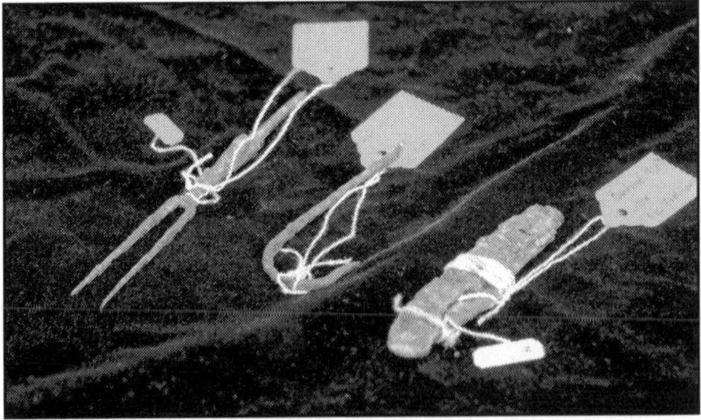

A fork, a fishhook, and a pocket knife, c. 1745, are housed at the North Adams Public Library. *[Mike Remillard Photos]*

A newspaper article dated November 13, 2000, indicates that North Adams Public Library officials discovered among the Fort's artifacts Elisha's spine with the musket ball still lodged in it and decided it would be appropriate to rebury it.

The decision was not without some local controversy. Though Elisha was originally scheduled to be interred on Memorial Day, some local historians thought it improper to bury a British colonial soldier alongside Revolutionary War veterans. The burial date was delayed.

After further discussion, historians came to the realization that Elisha did serve as a militiaman protecting and defending his state, then known as the Massachusetts Bay Colony, and died while acting in defense of Fort Massachusetts. It was agreed to bury him among other veterans. When Nims was killed in 1745, Massachusetts was a British colony, though he was defending the land that eventually became the City of North Adams.

Military ceremonies were conducted by veterans organizations at the Honor Roll Memorial outside the North Adams Public Library, where the American flag was presented to family representatives 255 years after Elisha's death. Burial followed on November 11, 2000, appropriately Veterans Day, at the Soldiers' Circle in Hillside Cemetery alongside some of North Adams' earliest settlers.

Postscript: *In 1763, at the end of the French and Indian War, France's dominance of the Americas ended and Britain came to control the land between the eastern coast and the Appalachian Mountains. The Treaty of Paris, signed in 1763, ceded all French claims east of the Mississippi River to the British (excluding New Orleans, which eventually went to Spain); in exchange, many lucrative islands of the West Indies were restored to France.*

It is estimated that over 25,000 American volunteers (militia), 34,000 British soldiers, and up to 1000 Native American allies fought against fewer than 6000 French soldiers, between 8000 and 11,000 Canadian militia, and 4000 Native American allies.

Casualties have been estimated at over 5000 for the British and 2000 for the French.

Ironically, in order to pay for the war, the British crown levied a series of taxes on their colonists. Those loyal to the crown considered this fair since the war had been fought on their behalf. Other colonists objected to the burdensome taxes and the lack of representation, which planted the seeds for the American Revolution. In 1775, only twelve years after the Treaty ending the French and Indian War was signed, revolution against the crown would commence.

Corporal John E. Atwood
Gettysburg Honor Guard
Civil War

"Four Score and seven years ago, our fathers brought forth on this continent a new nation." Corporal John E. Atwood, a native of North Adams, heard these now iconic words while attending President Lincoln's three-minute Gettysburg Address. It had been just eighty-seven years since America's war of independence, and another war was in full swing.

Corporal Atwood was present as a special guest, sitting just fifteen feet away from Lincoln. Atwood had survived being wounded in an earlier battle. He was convalescing from acute sunstroke and serving in the meantime as a nurse at a nearby hospital when he was selected as a member of the Honor Guard for the Gettysburg Address. It was a fine tribute to a man who, all in all, survived ten Civil War battles over three miserable years of warfare.

JOHN WAS BORN AT home in North Adams on October 9, 1839, the son of John K. and Sally (Jones) Atwood, and the third of their six children. His father John Sr. had been born in Rhode Island, while Sally was from Deerfield, Massachusetts.

There is little record of John's early life. It is assumed he attended local schools and, at some point, worked in the nearby mills as he did years later when returning from military service.

In September 1855, a group known as the Greylock Infantry was formed in North Adams; it later became known as the Johnson Grays, a local militia unit created in response to the increasing concern about Southern desire to secede from the Union and to prepare for the resulting possibility of war. The company was named in honor of Sylvander Johnson, chairman of the town committee. This industrialist oversaw the fundraising for the company to buy guns and obtain uniforms. The Grays formed a rallying point for the region's young men who responded to the call for arms during the ramp-up of what would indeed become a Civil War.

On April 18, 1861, a recruiting office opened, enlistment notices appeared in local papers, and 83 names were on the rolls within a week. John was one of the first enlistees. Interestingly, there is no mention of bounties for one's military service; initially, people were anxious and eager to serve. As the war took its toll, bounties would become more prevalent to encourage volunteers. Bounties could range from $100 or $150 up to even $700, with half provided up front and half when the soldier completed his service. Draftees did not receive a bounty, just soldiers' monthly pay: $13. Some less than honorable enlistees absconded with the first payment and never showed up at camp.

The Grays set up at Camp Johnson on the present grounds of the Massachusetts Museum of Contemporary Arts (on Marshall Street), then drilled and trained daily. They elected Elisha Smart to captain their company. Much needed to be done during the following weeks before traveling to Springfield to enlist formally with the 10th Massachusetts Regiment.

Sylvander Johnson obtained the cloth and the regulation cross buttons. Troops were measured, and uniforms were made in early June 1861 by the shop of Chapin & Biggs. The Ladies Aid Society made flannel shirts for the Grays in Burlingame and Ray's hall. Rifles were inspected in Boston, and troops were noted to be "parade ready" on Main Street in North Adams a week later with an "air of determination" (*Transcript* 6-20-1896).

John and the other troops were often accompanied on their drills and parades by the Hodges Cornet Band, whose performance inspired onlookers and audiences.

The company of Grays had strong local support. They were

The only known image of John E. Atwood. *[Courtesy of North Adams Historical Society Collection]*

feted with meals in South Adams (what is today Adams) and North Adams, had a large dinner on the lawn of the Congregational church with guest speakers, were presented with standards of colors, and responded by entertaining parade crowds. Their activities were reported weekly in the local newspapers. Enlistees came from the Berkshire-area towns of South Adams, North Adams, Williamstown, Monroe, Florida, Savoy, Cheshire, and Readsboro, Vermont.

John's parents were part of the "send-off" as the Grays left for Springfield, Massachusetts, to enlist into the Army officially; the group became Company B of the Tenth Regiment of the Massachusetts Volunteer Infantry on June 21, 1861. At that time, regiments were comprised of 1000 to 1100 men, with either ten or eleven 100-man companies.

They then traveled by train from Springfield to Boston and were rushed by steamships to Washington, DC, after the Union lost the First Battle of Bull Run and the capital felt threatened. In late July, the regiment arrived in Washington and were employed digging trenches and fortifying areas around the city. They commenced drilling exercises as well with their recently issued English Enfield rifles.

The Enfield 1853 musket-rifle was used by the North and South during the Civil War. It weighed nine and a half pounds, was a single-shot weapon, and usually could be fired about three or four times a minute. It was a .577 caliber and made a devastating or fatal wound if it hit a person in the chest; it often caused a limb amputation if the bullet hit a bone in the arm or leg. The rifle was highly sought after by poorly supplied Confederate forces.

Shortly after arriving in DC, the Tenth Regiment traded in their cadet gray uniforms and were issued the standard Union blue uniforms. John and his regiment wore hip-length, dark blue wool coats that fit loosely and hence were often called "sack" coats. In the summer, when it was hot, the men would rip out the worn, greasy coat linings to obtain some relief. Trousers were made of light blue wool, often with pockets. The men were issued leather ankle-high shoes called "brogans" that were often sucked off their feet in deep mud. Woolen shirts, socks, and underwear were also issued to soldiers, although most did not wear the warm underwear. Many men brought vests from home that provided additional warmth in the wintertime. Each soldier was issued the popular forage cap.

During the Tenth Regiment's time in Washington, from August through their first major battle in late spring some nine months later, they began to adjust, adapt, and prioritize what they would carry into battle. They would be walking hundreds of miles toting up to 50 pounds of gear, not including their rifle. John, like most soldiers, carried a knapsack on his shoulders that held items he did not need immediate access to, like an extra shirt, socks, a blanket rolled up on top with a half of a tent, a rubber blanket that could serve as a poncho or cover on the ground, an overcoat, and some cooking items.

He slung a haversack (catchall) over his shoulder, and it held the food he was going to eat that day, his pipe and tobacco, hardtack,

Bible, silverware, comb, sewing kit, pocketknife, a mug, dish, cup, hygiene items, newspaper, mirror, and playing cards. The haversack often carried food and thus became greasy. It had a removable liner that could be washed. Daily rations likely included a pound of salt pork, a potato, bread, coffee, hardtack, brown sugar, and some dried fruit or vegetables.

Over John's other shoulder, he carried his two-piece canteen.

The belt around his waist with a supporting sling around his shoulder supported a cartridge and cap box with ammunition and gunpowder for his rifle. The cartridge box held forty cartridges, each with a measured amount of gunpowder and a bullet wrapped individually in paper tubes. Any extra ammo John kept in his pockets or knapsack for easy access. He also carried his 21-inch bayonet in a leather scabbard on his belt.

As part of McClellan's Army of the Potomac, the Tenth Regiment and John's Company B began marching south in March of 1862, headed to their first major engagement. Finally, 100 miles south of Washington, closing in on the Confederate capital of Richmond, Virginia, they fought a pitched, face-to-face, bloody, two-day battle with the Confederate Army on May 31 and June 1, 1862.

The battle was known as Fair Oaks (or Seven Pines to Southerners), and the Tenth was at the center of the struggle. Overwhelmed several times, it rallied and pushed rebel forces back. The regiment and Company B fought bravely, and casualties were high. Casualties over two days of fighting, for both sides combined, totaled over 11,000 troops.

Company B's first battlefield losses occurred at Fair Oaks. The Commanding Officer, Colonel Briggs, was wounded in both legs, and a former Johnson Grays soldier from North Adams is credited for rescuing him from the battlefield. Captain Elisha Smart, the Grays' first elected captain, and three or four other Company B soldiers were killed in action. A witness is to have alleged that a Confederate soldier executed Captain Smart. John E. Atwood and other soldiers were wounded in this battle.

Records indicated Atwood was field hospitalized for several weeks recovering from an ankle or leg wound. It is presumed his wound was superficial and possibly bone-grazing, or through-and-through from either a mini ball or an artillery shard.

Though history indicates that the battle results were inconclusive, General McClellan claimed victory. However, he was so shaken by the experience that he is said to have written to his wife, "I am tired of the sickening sight of the battlefield, with its mangled corpses & poor suffering wounded! Victory has no charms for me when purchased at such a cost."

The Tenth Regiment and Atwood's Company B fought on with little respite. They were present at the exhausting Seven Days' Battles that began with fights at White Oak Swamp and continued at Gaines's Mill, Charles City Crossroads (also called the Battle of Glendale), and culminated in a Union victory at Malvern Hills on July 1.

For the remainder of 1862, the Tenth Regiment and Company B were refitted, involved in reconnaissance activities, and served in reserve and support at Antietam in September and Fredericksburg in December, suffering few if any casualties. Atwood, recognized for his abilities to direct privates in camp and on the battlefield, was promoted to corporal in December as the company went into winter quarters.

Coming out of winter headquarters in 1863 brought no letup in the fighting and dying. The Tenth Regiment was involved in the Chancellorsville Campaign, fighting at Salem Church and helping to capture Marye's Heights, suffering heavy losses. In early June 1863, the Tenth Regiment fought at Franklin's Crossing near Fredericksburg, Virginia.

Shortly afterward, the Confederacy decided to invade the northern territory and found its Army crossing the Potomac River and then, in late June, reaching the Susquehanna River in Pennsylvania. Ultimately converging on a small town called Gettysburg, the Confederate army, hoping to seize supplies but inadvertently clashing with Union troops, began on July 1, 1863, what would become a legendary battle.

Initially, when the Army of the Potomac, including the Tenth Regiment and Company B, heard of the Confederate Army's movement, they were many miles away near Fredericksburg. They began to march in sweltering temperatures to join what appeared to be a huge upcoming fight.

The Tenth force-marched through humid, blistering 80-degree heat, making the last 35 miles in 18 hours, enveloped in a thick cloud of dust, dressed from head to toe in heavy woolen uniforms. Some records indicate the march started at 2:00 a.m. and finished at 8:00 p.m. on July 2 when they reached Gettysburg. Upon arrival, the unit was directed to support the positions of the Round Tops, two rocky hill defensive positions.

By the end, the Tenth Regiment fielded 416 soldiers at Gettysburg; five were killed in an artillery barrage, and four were missing. Numerous soldiers, John Atwood being one of them, were felled by the heat: "As they were marching into battle at Gettysburg, he was overcome with sunstroke" (*Berkshires* 9-30-2017).

Atwood was taken to the field hospital with many others who were severely dehydrated. He was hospitalized and missed the battle. After his recovery, he remained at the hospital and served as a much-needed nurse, along with another Tenth soldier, Private William Mason. Gettysburg had been the costliest battle in US history and produced an estimated 51,000 casualties in three days. The hospital was overwhelmed.

Corporal Atwood continued to work as a nurse for months. He witnessed the aftermath of archaic battlefield surgery. While there was some medical advancement during the Civil War, the medical community in the 1860s was generally ignorant of the causes of diseases. Sterile dressings and antiseptic surgery were unknown. Doctors operated in primitive conditions, and while they did use chloroform, they continued to be unaware of sepsis and the source of infections. Medical treatments ranged from the ridiculousness of mud plasters and bleedings, to the fortunately sterilizing benefits of whiskey, to the more dangerous and debilitating dosing with opium. It is written that amputations took only 10 minutes, and surprisingly, 75% of the screaming, fever-ridden patients survived the ordeal.

Of the approximate 700,000 soldiers who died in the war, 400,000 perished from disease. The primary diseases were typhoid, diarrhea/dysentery, pneumonia, and malaria, many resulting from poor sanitation and hygiene, poor food and water, exposure, medical ignorance, and lack of clothing and shoes.

In late October 1863, Atwood was chosen by the governor of Massachusetts to be part of a three-member honor guard at the dedication ceremony of the Soldiers National Cemetery in Gettysburg, Pennsylvania. Corporal Atwood was selected to carry the Massachusetts state flag, accompanied by two riflemen. (In those days, when a soldier carried the flag into battle, it was better to die than allow the flag to be taken by the enemy or be dishonored by falling to the ground. The flag bearer is the most honored role of the color guard positions.)

Atwood could not have known that he would become a witness to one of history's greatest speeches. On November 19, 1863, in parade formation, Corporal Atwood, carrying the flag and accompanied by a rifleman on each side, marched to the cemetery with color guards from the other Union states. When reaching the cemetery, they mounted an immense platform that held several hundred people. Atwood was seated less than twenty feet from President Lincoln and heard every word of his famous Gettysburg Address.

Later that evening, after the dedication, Atwood had the further honor of shaking the President's hand at a reception.

Corporal Atwood is assumed to have continued his work at the hospital for the next eight months before honorably mustering out of the service in July 1864. He survived more than three years of active duty. While he continued to treat the wounded during those eight months, the Tenth Regiment fought in many well-recorded campaigns. There were skirmishes at Rappahannock Station, Mine Run, Brandy Station, and devastating battles at the Wilderness,

Spottsylvania, Cold Harbor, and Petersburg, where the Tenth lost many men before finally being ordered home, and the troops were themselves mustered out in July 1864.

It is notable that over the three years of active duty by the Johnson Grays (Company B), the war had a significant impact on its original roster. Forty-nine of them were killed or wounded in action, and thirty-four were discharged for sickness. Three had been taken prisoners of war. Mosby's infamous Confederate guerillas had captured William Harrington from North Adams; as the story goes, Harrington was able to get back to Union lines "with only his underclothing on" (*Transcript* 3-9-1920).

On December 22, 1865, the Tenth Regiment's Colors were brought by members of the regiment, presented at the State House in Boston, and permanently retired. In 1885, an eight-foot granite-based monument supporting a bronze sculpture of three muskets and a knapsack sitting on a drum was dedicated at Gettysburg to the 10th Massachusetts Voluntary Infantry Regiment.

Atwood, in his mid-twenties, returned home surprisingly unscathed, other than his ankle/leg wound, after taking part in ten battles over his three years of service. He took up residence with his parents and, seeking comradeship, joined the local G.A.R. organization named after Captain Charles D. Sanford. (Charles was a local man who had enlisted in the 27th Regiment and been killed in fierce fighting at Drewry's Bluff near the city of Petersburg, Virginia. There were over fifty other Northern Berkshire men in the unit, including Sanford's father, who was the chaplain. Father and son were buried at Hillside Cemetery in North Adams.)

Atwood, in his thirties, married Mary Holbrook, and they would have two children, a son named George and a daughter called Ann Elizabeth. Atwood juggled a number of local jobs and worked for years in a mill as a laborer, then as a printer, a carpenter, and a building mover. He was active in reunions of the Johnson Grays, the "Old Tenth," and the local C.D. Sanford G.A.R.

At fifty years old, in 1890, Atwood joined the North Adams Police Department. The local newspaper notes him walking a beat, investigating holdups, raiding illegal distilleries, and making arrests. The *Transcript* also stated he was a "man of courage and judgment and made an excellent officer" (10-3-1907). The article went on to report that "Mr. Atwood was a quiet and friendly man and very highly respected. He was a good citizen and a good neighbor, and it's safe to say had not an enemy in the world."

Another *Transcript* article, dated August 14, 1895, titled "Tramp Hunting—The Trying Work of our Police Officers on a Very Stormy Night—One Lone Catch," narrates Atwood and another policeman in the pouring rain, using a horse and buggy to conduct a successful midnight search at the fairgrounds to catch "dangerous" tramps. Their one "catch" is compliant and given a cell at the jailhouse.

By 1896, Atwood's health seemed to be deteriorating, and several friends who were "ex-policemen" were trying to help him secure a pension. They believed he might be entitled to a pension of $45 a month for his service in the "late war" (*Transcript* 4-17-1896).

Building records indicate that sometime in 1896, John E. Atwood built a two-tenement house on East Quincy Street for $3,000.

In 1897, Atwood applied for and was granted a liquor license for an establishment on Union Street, and later advertisements indicated he worked as an agent for Cavanough's, a local package store (*Transcript* 12-22-1902).

On May 17, 1904, John Atwood recited the Gettysburg Address as a special guest at the North Adams Memorial Day services. In a *Transcript* article, John was described as "a very quiet and modest man not given to exploiting his military experiences. For this reason, there are comparatively few people of this city who are aware of the part he took at the dedication of that great cemetery at Gettysburg, and undoubtedly the story will come as news to many of his comrades in the Sanford Post. He is in his 65th year but remarkedly well preserved and does not look over 50" (*Transcript* 5-17-1904).

In 1907, he had been working as a janitor when he traveled to

Boston for a second operation for throat cancer. He did not survive the surgery. John died at Massachusetts General Hospital in Boston on October 3, 1907, from cancer of the larynx at age sixty-seven (*Transcript* 5-31-1907).

John E. Atwood's body arrived in North Adams "on the 5:17 train" on October 4, 1907, and was "taken by Undertakers L.A. Simmons to the home of his sister. The funeral will be held on Saturday . . . ." The burial took place at the family plot in Hillside Cemetery, and the bearers were friends, members of the C.D. Sanford post, Grand Army of the Republic (G.A.R.), and the police force (*Transcript* 10-5-1907).

Sylvander Johnson, namesake of the Johnson Grays, is also buried in Hillside Cemetery.

In 1996, a commemorative ceremony was conducted at John Atwood's gravesite by the North Adams Historical Society and representatives of several veterans' groups. The program honored the memory of North Adams' contributions to the Civil War efforts. Nearly twenty descendants and a large group of city residents were present, including Tenth Regiment reenactors. The Gettysburg Address was read, and a G.A.R. marker made at a local foundry was placed on his grave, alongside the American flag. His great-granddaughter and great-grandson addressed the group, followed by a volley salute by the reenactors.

On Saturday, September 30, 2017, John E. Atwood was remembered once more when the Hillside Cemetery Restoration Commission conducted a ceremony, placing a new marker on Atwood's grave. The marker lists his military service, including being wounded in action at Fair Oaks, his police service, and his honor as a color bearer representing Massachusetts at the Gettysburg dedication. A chaplain read the Civil War soldier's prayer; the Gettysburg Address was again recited; an honor guard and salute were presented; and the police department placed the flag and new memorial holder at Atwood's stone marker. As a finale, "Taps" was played.

When the Grand Army of the Republic chapters were formed, many members were given blank Personal War Sketch forms to complete. The chapter members would often dictate to the chapter's secretary their experiences in the war—e.g., company and regiment, ranks they achieved, battles they were in, who their best friends were, and any experiences or memories they could share.

John E. Atwood's *Personal War Sketch*, recorded many years ago, is preserved and can still be seen at the North Adams Public Library. It was recorded by Atwood to the G.A.R. secretary as follows:

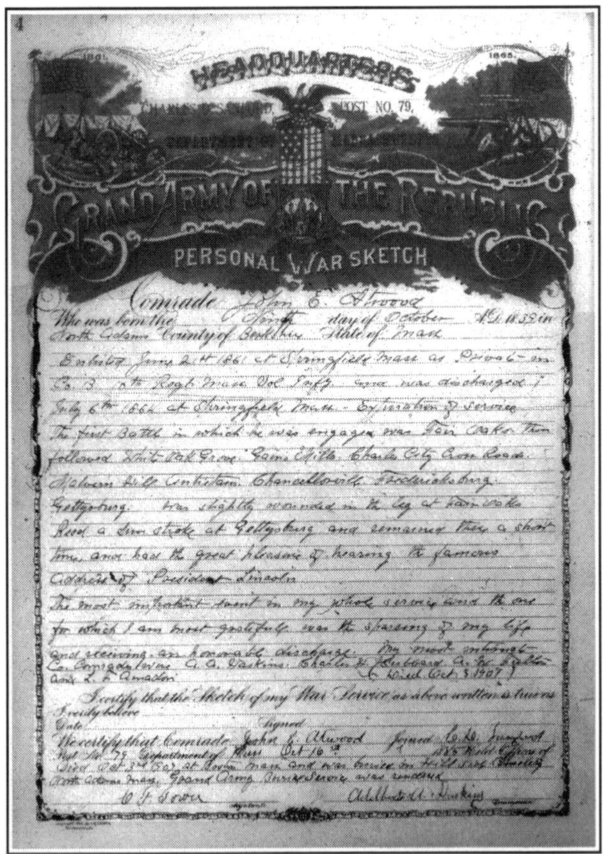
1861 HEADQUARTERS 1865
POST NO. 79,
GRAND ARMY OF THE REPUBLIC
PERSONAL WAR SKETCH

Comrade John E. Atwood
Who was born the [illegible] day of October A.D. 1839 in
North Adams County of Berkshire State of Mass

Enlisted June 21st 1861 at Springfield Mass as Private in
Co. B [illegible] Regt Mass Vol Infty and was discharged
July 6th 1864 at Springfield Mass - Expiration of service
The first Battle in which he was engaged was Fair Oaks then
followed [illegible] Oak Grove Gaines Mills Charles City Cross Roads
Malvern Hill Antietam Chancellorsville Fredericksburg
Gettysburg. Was slightly wounded in the leg at Fair Oaks
Recd a sun stroke at Gettysburg and remained there a short
time and had the great pleasure of hearing the famous
Address of President Lincoln
The most important event in my whole service and the one
for which I am most grateful was the sparing of my life
and receiving an honorable discharge. My most intimate
Co. Comrades were G. A. [illegible] Charles H. [illegible] [illegible]
and L. E. [illegible] (Died Oct. 3 1907)

I certify that the Sketch of my War Service as above written is true
[illegible]
Date Signed
We certify that Comrade John E. Atwood joined [illegible]
Post No. 79 Department of Mass Oct 16th [illegible]
Died Oct 3rd 1907 at [illegible] Mass and was buried in [illegible] Cemetery
North Adams Mass. Grand Army Burial Service was rendered
[illegible] Adjutant [illegible]

[Courtesy of North Adams Public Library]

HEADQUARTERS
CHARLES D. SANFORD, POST NO. 79,
GRAND ARMY OF THE REPUBLIC
PERSONAL WAR SKETCH

Comrade John E. Atwood

Who was born the Ninth day of October A.D. 1839 in North Adams, County of Berkshire, State of Mass.

Enlisted June 21st 1861 at Springfield Mass as Private in Co. B 10th Regt. Mass Vol. Infy. and was discharged July 6th 1864 at Springfield Mass. - Expiration of Service.

The first Battle in which he was engaged was Fair Oaks then followed White Oak Grove. Gains Mills. Charles City Cross Roads. Malvern Hill. Antietam. Chancellorsville. Fredericksburg. Gettysburg. Was slightly wounded in the leg at Fair Oaks. Rec'd a sun stroke at Gettysburg and remained there a short time and had the great blessing of hearing the famous Address of President Lincoln.

The most important event in my whole service and the one for which I am most gratefull was the spareing of my life and receiving an honorable discharge. My most intimate Co. Comrades were A.A. Haskins, Charles Hubbard, A.W. Fulton, and L.F. Amadon.

[Completed after his death] We certify that Comrade John E. Atwood joined C.D. Sanford Post 79 Department of Mass October 16th 1885. Died October 3, 1907, at Boston Mass, and was buried in Hill Side Cemetery North Adams Mass. Grand Army Burial Service was rendered.

[Witnessed by officers Tower and Adelbert A. Haskins.]

Corporal John E. Atwood, young warrior from North Adams, humble witness to Lincoln's Gettysburg Address, now rests with many of his comrades in Hillside Cemetery not far from where the Johnson Grays enlisted and trained for the War between the States. They could not have known what was ahead, but they carried on, preserving the Union through newfound courage and great loss.

Detail of John E. Atwood's grave marker. *[Mike Remillard Photos]*

Captain George L. Curran
Medical Doctor
World War I

This popular North Adams doctor was called to serve his country in the "War to End all Wars." A faithful healer, George Curran worked at a frenetic pace for hours on end, helping never-ending lines of wounded and sick patients, saving countless lives in a horrific, some say fruitless, war. He was thrown into a maelstrom that had already resulted in the deaths of millions from violence and a worldwide influenza pandemic.

BORN AT HOME ON Christmas Eve in 1887 to Dr. and Mrs. Charles and Catherine Curran, George was the second of seven children. His father, Charles, originally from Hadley Falls, was educated at Holy Cross College, attended Harvard, and graduated from the medical school of the University of New York. At one point, he met Catherine Lally from Williamstown and decided to relocate to North Adams upon his graduation in 1881. He and Catherine married in 1883.

Doctor Curran purchased a house on Eagle Street, across from St. Francis Church property and just south of the rectory. It became home as they grew to a family of nine, and it also housed the doctor's very successful medical practice. The children, especially his five boys, could observe first-hand the goings-on of their dad's vocation.

Three boys, including George, followed their father into the medical profession and became doctors. One brother, tragically, died at sixteen while attending Drury High School, though he too had planned to follow in his father's footsteps.

Doctor Curran was active in the community, serving on the North Adams Board of Health, and he was appointed to what was at that time referred to as the Massachusetts Board of Lunacy and Charity. He was also a delegate to the national Democratic convention for President Cleveland's second election.

It was a busy life for the large family. George received his primary education at St. Joseph's School right across the street, and then he entered Northside, a private preparatory school in Williamstown.

As a teenager, he was involved in a St. Joe's summer play production held at the Columbia Opera House under the direction of the Sisters of St. Joseph. George had great fun as one of the king's guards in *Sir Thomas More*.

He and his friends spent considerable time at the downtown YMCA. George was an active member of the gymnastics team and participated in the group's annual public carnival event exhibitions, featuring dumbbell drills, mat exercises, horizontal bars, and the pommel horse, George's specialty. According to the local newspaper, the event caused "considerable merriment" among the spectators.

He decided to attend Amherst College as a freshman from 1905 to 1906 and played left field on their baseball team, then summer ball for a local team in Cobleskill, NY. Soon after, George entered Bellevue Medical School of New York University as a member of the class of 1910.

Studious and bright, George received word in August 1910 that he had successfully passed the State Board Medical Exam for Massachusetts, after four years of study. The news article noted that "Massachusetts and Colorado are the only two States in which applicants for the state board are allowed to take their examinations before receiving their degrees from their universities. . . . The fact that Mr. Curran still has another year to complete at the New York University School of Medicine is all the more reason for congratulations" (*Transcript* 8-12-1910).

Upon graduation, George came home and joined his father's medical practice on Eagle Street. Sadly, his father, who had served the local community for thirty years, was in declining health. Four months after George arrived home, Charles passed away at the age of fifty. George's mom, Catherine, had passed away in 1908.

George assumed his father's thriving, well-established medical practice on Eagle Street and took up residence there, helping to support his younger brothers and sisters remaining at home.

George enjoyed just three years of doctoring in his hometown before World War I began on July 28, 1914, triggered by the assassination of Archduke Franz Ferdinand of Austria.

For many years, tensions had been simmering in southern Europe. Initially, the United States assumed a position of neutrality that was abandoned after Germany's use of unrestricted submarine warfare, which sank the British ocean liner *Lusitania*, killing over 100 American passengers. In the ensuing months, Germany continued sinking American vessels. Their actions swayed American public opinion away from isolationism, and the US declared war on Germany on April 6, 1917.

At the time, the United States had an army totaling 200,000 men, 80,000 of whom were National Guardsmen. They needed millions, quickly. A month after the declaration of war, President Wilson signed the Selective Service Act requiring all men between the ages of 21 and 30 to register with the government, instituting the draft and beginning the conscription process for troops. By the end of the war, 4,700,000 soldiers would serve, 2,700,000 of them overseas, most of whom were drafted.

At the age of thirty-one, Doctor Curran saw the impending needs of his country, paused from his busy practice, and applied for admission to the United States Army Medical Corps. The process was speedy, and on February 14, 1918, he was accepted as a first lieutenant into the medical corps reserve.

All three Curran brothers joined up, as documented in the *Transcript*: "His brother Lieutenant Arthur M. Curran . . . is in France as a surgeon of the 104th Regiment. . . . [T]he third brother, William Curran, who was a medical student, is also in the medical reserve corps" (2-14-1918).

A month later, George received orders assigning him to Walter Reed Hospital in Washington, DC. Due to his years of experience, he was drawn to the increasing number of unusual cases that the hospital received from surrounding military bases. The illnesses often resulted in pneumonia and other respiratory problems.

At the time, the influenza pandemic of 1918 was beginning, and in retrospect, many think it started, or at least had thrived, on crowded military bases. Doctor Curran did not know what type of disease he was facing, marked by sudden and high fevers, dry cough, headaches, chills, fatigue, etc.

In 1918, there was no effective treatment. No medications, no antibiotics, and no ventilators existed to support soldiers long enough for their immune systems to vanquish this mystery virus. Cramped military quarters could only exacerbate the situation, and the rush to transfer men overseas to fight helped disperse the flu worldwide.

George remained at Walter Reed for about four months before shipping out for France and arriving mid-August 1918.

Photograph of the highly esteemed Captain Curran. *[Reprint from the* Transcript, *June 30, 1938]*

"Lieutenant George Curran is now on the Other Side," read the headline in the *Transcript*. "[He] has arrived safely overseas. He is now stationed at a base hospital in France" (8-14-1918).

George was immediately assigned to the staff at Base Hospital 50 of the American Expeditionary Force at Mesves, near Bulcy, just a few kilometres from the Loire River. (By comparison, hospitals around Paris were typically 12 hours from the front.) There he cared for the burgeoning number of pneumonia and influenza cases and hundreds of

Base Hospital 50 camp, with tents in background, Mesves, France, approximately 1918-1919. *[University of Washington Libraries, Special Collections, SOC10457]*

casualties brought in from the casualty clearing stations. It was noted that at the peak of his time at the hospital, Dr. Curran "was carrying a load of 500 patients" . . . and he returned home "from France nearly 50 pounds lighter than when he went" (*Transcript* 6-30-1938).

An old black and white photograph of Base Hospital 50 shows a worn sprawling campsite with numerous tin-roofed casualty huts, muddy walkways, and countless tents in the background. In the foreground is a rail siding, or spur, meant to move casualties to more sophisticated treatment. The camp was authorized by the US War Department for the American Red Cross in October 1917 under the auspices of the University of Washington. It remained in France from August 1918 to February 1919.

When the United States entered the war, Congress authorized the American Red Cross to organize fifty base hospitals and recruit nurses from universities and civilian hospitals. Hospital staffing usually included 23 medical officers, 60 nurses, and a complement of N.C.O.s and enlisted men.

The United States spent most of 1917 and the early part of 1918 conscripting and training Army soldiers and transporting them overseas. Once in Europe, they trained months with British and French units who had been at war for years before taking over responsibility for sections of the Allied trench lines. It wasn't until Spring 1918 that most units were ready for front-line service.

Medical Profession—World War I

By the time the Americans entered the war, almost three years had passed since its beginning in July 1914, and millions of lives had been lost. The medical profession had been ill-prepared for the violence, scope, and volume of carnage that the war would produce. By the time Dr. Curran arrived in August 1918, well over 5,000,000 soldiers had died and millions more were severely wounded.

Few medical professionals understood the effects of 20th-century weaponry on the human body. High explosive artillery produced flesh-tearing shrapnel; machine guns riddled and tore bone and muscle; and gas blinded and asphyxiated millions of soldiers. Medical science was not up to date on how to treat these vicious new ways to harm people. To make matters worse, the fertile, manured fields of Europe worsened bacteria-laden wounds, leading to quick infection.

Doctors also dealt with the blistering effects of mustard, phosgene, and chlorine gases with no known antidotes. (Phosgene was uniquely diabolic: "At low concentrations, its odor is similar to that of green corn or new mown hay," says the CDC, meaning it could blend right into pastoral scenery just before turning sharp and unforgiving.) Even those lucky enough to avoid the gases lived with trench foot, fever, body lice, dysentery, and typhus, everyday maladies experienced by men huddling for months at a time in any of the hundreds of miles of muddy trenches.

Centuries-old military tactics that focused on frontal assaults ensured tens of thousands of casualties. Two Maxim machine guns, each spewing out 500 hundred rounds a minute at velocities of

3000 feet per second, could pulverize bones and muscle and decimate a charging 1000-man infantry battalion in a matter of seconds, leaving hundreds of soldiers slumped in a decaying wasteland. It was a volume no one expected. In four days of combat in August 1914, French forces recorded 140,000 casualties.

Medical thought and practice in the early stages of the war were deeply archaic. In 1914, some of France's most prominent physicians subscribed to and promoted the accepted theory that gunshot wounds should not be probed, because that type of action would infect the site. It was maintained that the wound would heal on its own and needed minimal care. This mentality led to many infections, a common malady called gas gangrene, and untold deaths, with lucky survivors requiring agonizing amputations.

Beyond the lack of evidence-based medicine, a simpler practice was at fault: the delay in treating the wounded. Stretcher-bearers had been taught poor protocol. When wounded men were brought off the battlefield, there was an initial lack of triage and then interminable delays in transporting the wounded to the rear; after some time and many deaths, a realization set in that triage and treatment needed to occur closer to the front.

After three years of wholesale slaughter, the war began serving effectively as a bloody laboratory for medical professionals and technicians to solve answers to some of the worst suffering ever witnessed. Medical science advanced at great cost.

By the time Dr. Curran arrived at Base 50, the medical corps had determined that the wounded should be triaged and operated on as quickly as possible. The seriousness of the wound should have a direct impact on the length of time it took to evacuate the wounded soldier. Supporting this effort, a dedicated stretcher-bearer corps was created and trained in hygiene, advanced first aid, and controlling hemorrhages.

The armies set up better equipped Casualty Clearing Stations (CCS) closer to the battlefield, fully staffed with doctors and nurses, allowing for more immediate intervention. In the past, the wounded soldier would be shuffled by ambulance, screaming, over deeply rutted roads to various aid stations, then to a train which would often be delayed. Ammunition and resupply trains took priority.

The Great War spawned several new medical fields; one was radiology. Marie Curie enthusiastically adopted the use of X-rays previously discovered by Wilhelm Röntgen. She outfitted a number of "radiologic" Renault vans that traveled to the front lines in 1916. She became so involved she got a driver's license and drove herself to the front, sporting a Red Cross armband.

The vans became highly desired; hospitals across the front called for the mobile units. The soldiers called them *P'tites Curies* (Little Curies). Before X-ray machines, surgeons would take out what shrapnel they could see and wait for infections to arise to remove more. The *P'tites Curies* would identify the location of shrapnel before infection could set in.

Neurology was another new field. The war produced thousands of head injuries that physicians were reluctant to deal with in the past. Most wounded either died from infection or from the clumsy intervention of drilling burr holes in the patient's head (a technique still used today to great success, but under more controlled conditions). Outcomes improved dramatically with newly introduced techniques, quick actions, and cleanliness.

Better use of transfusions, antiseptics, and morphine helped reduce casualty fatalities. Another significant introduction was the Thomas splint, which immobilized shattered appendages, replacing bulky ineffective contraptions. Invented in 1875 by a Welsh physician of the same name, its application was championed during wartime by Thomas's nephew, who happened to be a consultant orthopedic surgeon to the British Army. By the end of 1915, the splint was mandated at all base hospitals. It was a godsend; previously, half of the wounded died from a broken femur due to gunshot wounds.

By 1918, the Field Medical System had been greatly improved, although thousands of casualties continued to arrive daily.

When George was stationed at Base Hospital 50, the Allies had begun the Battle of Amiens on August 8, 1918, also known as the

Hundred Day Offensive. For the next 100 days, starting in Amiens, the Allies, with fresh American troops, in a concerted effort, started an advance that dislodged and demoralized German forces. This operation ultimately led to the end of World War I.

On the other hand, this massive offensive created tens of thousands of casualties that, after being stabilized at a CCS, were quickly sent by train back to hospitals, including Doctor Curran's south of Paris.

A new wave of influenza was also arriving. It's worth noting that many doughboys, as the American infantrymen were called, were already infected with the flu as they walked down gangplanks in France several months before the massing for August's offensive. Some were hospitalized, but most headed to the trenches where cases skyrocketed in the miserable conditions: fetid water, shell holes with feces, rotting dead men, and rapacious rats . . . a perfect environment for viral spread.

Hospitals were desperate. Bloodletting was even tried. Morphine seemed to help with the coughing, but thousands were dying, more if counting pneumonia deaths.

Influenza spiked during the United States' greatest campaign of the war. From September–November 1918, the flu sickened 20 to 40% of all US Army and Navy personnel. At one point, it was said that influenza cases outnumbered the Allied combat casualties. One tally had 227,000 hospitalized for wounds and 340,000 for flu.

Some months after his arrival at Base Hospital 50, George and other medical officers attended a brief lecture on traumatic shock at Base Hospital 103 in Dijon, by Doctor Walter Cannon, a noted authority. At the time, Doctor Cannon had determined that, in many cases, blood transfusions seemed to lessen the deadly effects of shock. Doctor Cannon was at the forefront of understanding how to deal with traumatic shock dominating so many battlefield cases.

Toward the end of his overseas tour, Doctor Curran was transferred to Dijon as the battlefield needs subsided. For someone who'd had little rest over many months, this was surely a relief.

A retrospective news article in the *Transcript* notes:

> *After months at Base Hospital 50, Dr. Curran was transferred to Base Hospital 103 at Dijon. There he was on duty in the Spring of 1919 when he was summoned to Paris to receive from the hands of the French Minister of War, the medal of honor awarded him by the French government in recognition of his service to the troops of its Army who came under his care in the hospitals in which he served (6-30-1938).*
>
> *THE CITATION:*
> *In the name of the French Republic*
> *The Minister of War has awarded the*
> *Silver Medal of Honor*
> *To First Lieutenant*
> *Doctor George L. Curran*
> *of the Hospital Center at Mesves.*

To the French, Lieutenant Curran gave enlightened and devoted care. He was hereby authorized to wear this medal in his coat lapel suspended by a tricolored ribbon. The certificate was given in order to perpetuate among his family and his co-citizens the remembrance of his honorable and courageous service.

Doctor George Curran served with the American Expeditionary Force (A.E.F.) in combat from August 8, 1918, through July 1919, eight months beyond Armistice Day (November 11, 1918), helping those grievously wounded and sick.

By the end of the war, the United States had lost 116,708 soldiers: 53,000 to combat and 63,000 to disease (predominantly influenza).

The total deaths across all armies were estimated to be 8.5 million. Some say the carnage of those blood-soaked battlefields allowed medical science to advance in four years what would have taken a generation under normal circumstances.

A *Transcript* article headlined on July 8, 1919:

> ***Captain G. L. Curran Arrives in New York****—Word was received here today that Captain George L. Curran, a former resident of this city, had arrived in New York City from base hospital No. 103 in France. The trip across has been made on the* Great Northern. . . . *Dr. Curran was sent overseas for about a year, and he has been stationed in various base hospitals in France. On March 18 of this year, he was given the French military medal because of his tireless attention to duty in the face of the greatest of obstacles. This decoration was conferred upon Captain Curran in Paris.*

Shortly after his arrival on July 8, Doctor Curran was granted a furlough, and while on leave, he married Claire E. Russell, a registered nurse formerly from the state of Washington, a member of the first group of Red Cross nurses who were sent to France in 1917 and assigned to Base Hospital 50. It was there that she met George.

The young couple was wed on July 21, 1919, at St. Ambrose Church in Chicago, and George returned to New York City to be honorably discharged on July 25, 1919.

The couple settled in George's hometown of North Adams, becoming enthusiastically involved in the community. George served three years as president of the Northern Berkshire Medical Society and at one point was president of the medical and surgical staff at North Adams Hospital. He returned to the thriving practice he'd left seventeen months before.

Claire became a member of the school committee and the library board. She also belonged to the V.F.W. Auxiliary and the Travel Club. The Currans went on to have two sons; both became doctors.

In 1921, the three Curran brothers, George, Arthur, and William, all physicians, veterans, and most importantly, survivors of The Great War, decided to open the Charles L. Curran Memorial Clinic, named after their father (*Transcript* 7-27-1921).

This state-of-the-art clinic resided in the family's former residence on 63 Eagle Street. The converted space housed a "modern medical and surgical institution," including "an exceptional laboratory, an x-ray apparatus, an operating room . . . modern appliances for diagnosis." There would also be a technician in attendance, a nurse available "day and night," "24-hour telephone service," and "Dr. George Curran would specialize in obstetrics and gynecology while his brothers would attend to abdominal surgery and children's cases."

Civically, Dr. Curran spent considerable time working with veterans, the V.F.W., and the American Legion. He became the local V.F.W.'s 11th Post Commander and devoted much time to reviving its flagging membership. (The post would one day be renamed in his honor.)

News articles from the period show the doctor responding to calls for auto accidents, births, industrial accidents, athletic injuries, mishaps, and certifying deaths; in one interesting incident, Doctor Curran was called to the Berkshire Hotel to "dress an injury" when a salesman "sustained a severe cut on the head when he fell down the stairs . . . after which he was taken to the police station to be cared for because of his condition." The newspaper notes, "he was formally booked on a charge of drunkenness . . . and is in an extremely nervous condition which is not attributed to the fall" (*Transcript* 7-7-1924).

Apparently, the demands of his thriving practice were affecting his health. The *Transcript* notes in September 1927 that Dr. Curran had returned to his "medical practice following a three-month period of illness and convalescence from pneumonia."

During the same year, the Curran family was also shocked when his brother Arthur Curran died at age thirty-five from what was described as a "shattered health condition" attributed to his time as a surgeon during his front-line service with the 104th Infantry Regiment in World War I, and before that, his deployment to the Mexican border.

Years later, in honor of his service to country and community, North Adams dedicated its newly completed link in the city's main highway to the late Dr. Arthur M. Curran. It's now known as Curran Highway.

(Reprinted from The Transcript of Thursday, June 30)

Dr. Geo. L. Curran Dies Suddenly; Community Shocked And Saddened

End Comes Few Hours After Shock — Funeral at St. Francis Sat. at 10.30 a. m.—Body to Lie in State at Clinic Before Services — Premonition of Death Seen in Fact He Had Named Pall Bearers—City Hall Flag at Half Mast—Friend of Sick and Unfortunate Widely Mourned.

Dr. George Lally Curran, physician, surgeon and first citizen of the community in the hearts of hundreds of its people, is dead.

He passed away in his 51st year at 4 o'clock this morning at his home, 170 Pleasant street, from the effects of a cerebral hemorrhage he suffered at 8 o'clock last evening in his office at the Charles J. Curran Memorial clinic on Eagle street.

He had ministered to the first of the usual large group of patients who during the early evening hours had gathered in his waiting room, and was about to receive the second at his consultation desk when he suddenly slumped over in his chair. Miss Christine Teitsch, the clinic nurse and his office assistant, was in the room with him and she immediately summoned Dr. James W. Bunce, who was at his side within a few moments.

Sinks Steadily Until End

Dr. Bunce found Dr. Curran in a state of partial unconsciousness, and recognizing instantly the symptoms of a severe shock, ordered his immediate removal in the ambulance to his home. There Dr. Bunce and Dr. William A. Nelson of Williamstown, who had worked with Dr. Curran on scores of cases during their parallel careers in the neighboring communities, remained at his bedside until the end, laboring in a vain effort to stay

[Transcript, June 30, 1938]

For the next decade, the clinic continued to treat thousands of patients until one day in the early evening hours of June 29, 1938, after already seeing dozens of patients and preparing to see more, Doctor George Curran slumped over his desk. He was diagnosed as the victim of a cerebral hemorrhage, moved to his home, and passed away early the next morning.

His family decided that the beloved doctor would lie in state at his clinic and former home, where friends and associates could pay their last respects. The flag at City Hall was lowered to half-mast, then the doctor's remains were brought to St. Francis Church for a funeral Mass and interred at St. Joseph's Cemetery.

The doctor requested no special military honors, although all his bearers were World War I veterans. Thousands came to pay tribute, and Dr. George Curran's final goodbye became one of the city's most well attended funerals.

The community mourned the loss of its cherished citizen, its hometown healer. The country lost a patriot. As recounted by the *Transcript* on June 30, 1938, the mayor of North Adams, in a final tribute to Curran, said: "His life was one of real service and personal sacrifice. There is hardly a home to which he has not brought solace and comfort. His friendly way and kind word brightened the heart of many."

Admiral Allan R. McCann

Submariner

World War I and II

In May 1939, thirty-three submariners were rescued from the depths of the Atlantic using a bell-shaped McCann Rescue Chamber. For the first time in history, men were saved from a disabled, sunken submarine resting precariously on the bottom of the ocean at a depth of 240 feet. The development of the McCann Rescue Chamber was one of the pinnacles of Admiral McCann's achievements over his three decades of illustrious naval service that spanned two World Wars.

JAMES MCCANN, FATHER OF Allan, was of Scottish background and emigrated from Canada to the United States in the 1880s. Initially, James worked in New York, then in 1887 he moved to North Adams. James met and married Carrie Utman in 1896, a young lady from Blackinton. Interestingly, both were involved in the apparel business. Before their marriage, Carrie had a dress-making business, and James worked approximately twelve years as a head cutter for P.J. Boland's, a high-end tailoring establishment.

In 1906, James became a partner and proprietor of the P.J. Boland Company—Tailors and Haberdashers—with stores in North Adams and New York. The North Adams business was located on Bank Street in a redesigned building occupying three floors.

James's brother William came to North Adams about the same time and worked with him at P.J. Boland. In 1925, James and Carrie returned to New York, while William remained in North Adams, becoming a partner in Smith and McCann's tailoring firm. Tragedy struck when his body was discovered by the janitor in the Kimball building on September 16, 1928, having succumbed to a gas leak (*Transcript* 9-17-28).

James and Carrie had two sons, one of whom became the future Navy Vice Admiral Allan Rockwell McCann. Allan was born in North Adams on September 20, 1896. He attended Mark Hopkins School, then Drury High School. Early on, he was nicknamed Mack, and as a boy he enjoyed fishing, hunting, and sledding with his friends and, like so many residents, swimming at the local Windsor Lake. Allan also took an interest in city government and served as a page for the city council.

While at Drury, Allan received his appointment to the Naval Academy and left before the end of his sophomore year to study for the entrance exam at a preparatory school in Annapolis. He passed the exam in June 1913 and was formally accepted to the Naval Academy at 16 years of age.

Allan studiously applied himself, yet graduated in the lower half of his class. The Navy was in dramatic need of officers. He was commissioned an Ensign in March 1917, several months earlier than the regular June graduation. In April of 1917, war would be declared on Germany.

Assignments came quickly. Allan served in the Atlantic Fleet as the Watch and Division Officer aboard the battleship USS *Kansas*. During World War I, the ship transported soldiers to and from France. When Allan was assigned, the *Kansas* was often used as a training ship for Naval Academy cadets and graduates, their first introduction to the fleet.

Attracted to the submarine service, Ensign McCann found himself assigned to the recently established Submarine School in New London, Connecticut. The billet dovetailed nicely with his mechanical aptitude. He came to the realization that underwater vessels would surely play a large part in the Navy's future and were likely to be the quickest avenue for him to command a ship.

Assignment to the school also placed him closer to North Adams, where his fiancée, Katheryne Gallup, was residing. The young couple married at St. John's Episcopal Church in what was described as a "very quiet" ceremony on October 7, 1918. Katheryne, described in the *Transcript* as "one of the city's popular young women," was the daughter of well-known Harvey Gallup, a former senator, mayor, and local insurance businessman whose United States ancestry stretched back to the 1600s. Over the next few years, Allan and Katheryne were blessed with three daughters.

The young couple returned to New London, and Allan resumed his four months of daunting studies in submarine mechanical, pneumatic, and electrical systems. He also received considerable training on submarine weaponry.

The challenging school was intended to prepare officers to work closely with their crew under stressful underwater conditions. During the early days of submarines, there were many physical risks, often characterized by subs sinking from unexplained explosions, accidents, or unanticipated flooding. Sometimes called the "iron coffin" service, submarines were not for everyone, and those who did not complete their training were returned to the surface fleet.

In the 1920s, Allan, now Lieutenant McCann, was assigned to different submarines and, in 1922, became a division commander and in charge of a submarine that patrolled the Panama Canal. He served in the Canal Zone several times, protecting US interests.

The 1920s was also a time of submarine disasters across the globe. Several subs were sunk in US waters when hit by other vessels or due to mechanical mishaps. Up to this point, the only possibility in saving the crew was trying to hoist a sub to the surface. Seldom did this effort produce good results; in most cases, the crew still drowned or were suffocated.

A US submarine disaster in 1927 yielding the loss of forty men created a public outcry and prompted the Navy to search for a device to rescue trapped sailors. Lieutenant McCann began thinking there must be a better way to save submariners than waiting days for a hoist to arrive.

He was briefly distracted from his thoughts in 1927 when three submarines he led served as lifeguards for the Dole Air Race (Dole Derby) when pilots competed in a 2400-mile race from San Francisco to Hawaii. His submarines were charged to search for downed pilots. Several planes did not complete the race, although none was ever recovered.

In 1928, Lieutenant McCann was assigned to the Navy's Bureau of Construction and Repair, where he and a Lieutenant named Momsen had separately begun sketching out plans for a submarine rescue chamber. Momsen is credited with the original design submitted to the Navy but initially ignored by the bureaucracy.

A rescue diving chamber on the deck of the USS Falcon during the early stages of the USS Squalus salvage operations in May 1939.

McCann was put in charge of revisions to Momsen's "diving bell." Momsen's design showed a large bell with an open bottom, cables to attach it to a sunken submarine, and a water-tight rubber gasket to suction the bell to the submarine. The two-chamber diving bell carried rescuers in the top portion and once attached to a sunken submarine, allowed evacuees into the bottom part. The completed device was unveiled in 1931, and it became known as the "McCann Rescue Chamber." The United States gave the plans to thirteen other nations, and McCann rescue chambers were distributed at key naval bases in New London, Hawaii, San Diego, Key West, and Cavite in the Philippines.

The chamber was described as "a miniature submarine, except most of its functions are supplied by outside sources" (*Transcript* 5-25-1939). The pear-shaped cylinder weighed 18,000 pounds and was ten feet tall by eight feet wide. The engineering seems simple at first glance, but at depth, every step is complex: Three operators of the two-chambered bell would enter the sealed top chamber, supplied with air by the salvage vessel. The chamber would be lowered along a cable that has been attached by a diver to the sunken sub's hatch. Ballast tanks ensured the bell remained upright and controlled its buoyancy on the descent.

Once the chamber reached the sub, its five-foot opening (on the bottom of the lower chamber) with a huge rubber gasket would be suctioned against the sub's shell, and water would be expelled from the lower chamber. When the hatch between the two chambers was opened, the sub's hatch was opened, and six to eight people could move into the lower chamber. Once the rescued were in the lower chamber, the sub's hatch was sealed, the bottom hatch in the lower chamber was closed, and the bell was winched up to the salvage ship.

The Bureau of Construction and Repairs awarded McCann and others commendations for developing the submarine rescue chamber. The commendation read: "that through your intelligent application of an intimate knowledge of submarine material and submarine operation, and the various problems with the deep-sea diving and through great personal risk in subjecting yourself to the dangers of underwater submarine abandonment while in the experimental stage, you have contributed directly and in large measure to the successful development of means for rescuing entrapped personnel from sunken submarines."

In 1931, the Navy became interested in how submarines would function under the ice cap and decided to reconfigure a submarine to travel from the Atlantic to the Pacific Ocean via the North Pole. McCann, now at the rank of Lieutenant Commander, was a reliable and solid structural engineer. He supervised the thirty modifications, outside and inside, to a submarine that was to be called the *Nautilus*.

The *Nautilus* failed in its attempt to sail beneath the Arctic ice cap. However, much was learned, and Commander McCann gained experience in his unique efforts of attaching an ice drill, a telescoping conning tower, a diving bell, and an air lock to the reconfigured sub.

During the mid-1930s, McCann was assigned as the skipper to the *Bonita*, a V-Class submarine, and served on several cruisers, including as the damage control officer to the USS *Indianapolis*. Back home, his father had grown unwell and passed away in 1936. His mother, Carrie, followed in 1941.

Not long after, Lieutenant Commander McCann was assigned to the cruiser USS *Indianapolis* for a second time. Under this assignment, he served as third in command charged with carrying President Franklin Delano Roosevelt to Buenos Aires for the Pan American Peace Conference on December 1, 1936. The *Indianapolis* was 610 feet long, armed with 9 eight-inch guns, 4 five-inch anti-aircraft guns, and 2 three-pounder saluting guns. It also carried 4 scouting seaplanes and was capable of a speed of 33 knots (*Transcript* 11-19-36).

NOTE: In July 1945, while returning to the United States after delivering parts for the atomic bomb to a base in the Pacific, the USS *Indianapolis* was torpedoed by a Japanese submarine. The cruiser sank in minutes, and of the 1195 crew members on board, only 316 survived. It was the greatest loss of life from a single ship in US history, and it was just weeks before the Japanese government surrendered. In 2017, the wreckage was located in 18,000 feet of water. In 2018, the crew of the *Indianapolis* was collectively awarded a Congressional Gold Medal.

On January 28, 1937, a North Adams newspaper noted that Lieutenant Commander McCann was "promoted to Commander (Army equivalent of a Lieutenant Colonel) and that Commanders are usually given command of submarines or destroyers." The article also mentions the salary of Commanders: $5,275 annually, plus allowances.

It was only a matter of time before the next mechanical malfunction would occur on a submarine, and so it did in May 1939 when the submarine USS *Squalus* and its crew of 59 sailors sank in over 200 feet of water off Portsmouth, New Hampshire. Commander McCann was made a technical aide and member of the "*Squalus* Salvage Unit" assigned to the rescue unit. More than a score of experts was flown to the scene and met the support vessel USS *Falcon*, which carried the rescue bell from New London.

The rescue party did not know that 26 men had already perished inside the submarine, but the *Falcon* quickly lowered and attached the McCann Rescue Chamber to the *Squalus*. Two operators manned the upper chamber and, over four different trips, extricated the remaining thirty-three crew members, a feat that had never occurred in such a depth. The *Squalus* was eventually hoisted from the depths and the bodies recovered. Most of the rescued sailors continued in the submarine service.

McCann "always insisted modestly that the diving bell was a 'bureau development' in perfecting the mechanism, and that other officers and men had had an important part as himself. However, his associates in the service maintained that the principal contributions to its development were his and gave it the name 'McCann Bell'" (*Transcript* 1-28-1937).

McCann and seven others personally received from the Acting Secretary of the Navy a commendation for their efforts from President Franklin Roosevelt, noting, "Your determined and efficient efforts have held the attention of the entire nation and the successful completion of this unprecedented task merits the highest approval and admiration . . . . Well Done!!"

The Navy built four rescue chambers in addition to the one used to rescue the *Squalus* submariners. They were based in the Panama Canal Zone, on the West Coast, at Honolulu, and in the Far East. Newer versions of rescue chambers now exist but employ many of the same characteristics of the McCann Bell, which for decades provided hope that rescues were possible.

McCann's persistent efforts surrounding the rescue chamber helped establish his reputation as a hard-driving military man with an inquisitive nature who fostered an interest in science. Later efforts would establish him as a brilliant tactician involving undersea and surface ships.

NOTE: The recovered, repaired, and recommissioned *Squalus* was renamed the USS *Sailfish*. It distinguished itself during World War II, completing numerous Pacific patrols, damaging and sinking numerous enemy vessels. The *Sailfish* was awarded nine battle stars and received the Presidential Unit Citation for successful "operations against the enemy in Japanese-controlled waters."

Not long after the *Squalus* recovery, Commander McCann left the Portsmouth Naval Yard as Commander of Submarine Squadron 6, Pacific Fleet in Hawaii. His family remained in DC, and McCann arrived in Pearl Harbor in time to witness the Japanese attack on December 7, 1941. From his sub tender USS *Pelias*, McCann directed gunnery counterattacks against the Japanese and participated in rescue efforts. Luckily, his twelve submarines survived unscathed and were heavily deployed, fighting the Japanese Navy for the next several years.

McCann also helped resolve an ongoing problem with torpedo failures that endangered his submarines during this time. After months of missing many opportunities to sink Japanese ships, through multiple inquiries and tests, it was determined that torpedo firing pins were defective, and the problem was corrected.

Promoted to Captain in 1942, McCann was transferred to command Sub Squadron 7 in the Atlantic. He was now in battle, directing operations against Germany's U-boats. Captain McCann supervised the development of common tactics to fight Germany's wolf packs (groups of German submarines intent on destroying Allied shipping). His efforts included establishing uniform tactics that helped destroyers and their escorts find and destroy U-boats. His efforts, alongside those of many others, saw a dramatic increase in the monthly number of U-boats sunk.

Captain Allan R. McCann, USN Photographed in Service Dress Grey Uniform, circa 1943-44, while he was serving with the Office of the Chief of Naval Operations, in Washington, D.C. *[Official US Navy Photograph, now in the collections of the National Archives. Under Public Domain]*

Captain McCann was transferred back to the Pacific Fleet in 1944 to command the battleship USS *Iowa,* which supported landings in the South Pacific, including Peleliu and the Philippines. During this one-year assignment, McCann was promoted to Rear Admiral and awarded the Bronze Star for his achievements. McCann was now in charge of over 2500 sailors and a ship more than 887 feet long with powerful sixteen-inch guns that could shoot 2700-pound armor-piercing shells over twenty miles. Launched in 1940, the *Iowa,* known as the "Big Stick," garnered nine World War II battle stars and a Navy Meritorious Unit Citation.

In 1944, Captain McCann was awarded the Legion of Merit for his meritorious service as a submarine squadron commander in the Southwest Pacific during World War II. Part of the award citation reads: "Captain McCann demonstrated exceptional ability and untiring devotion, inspiring officers and men under his command."

Transferred again in December 1944, McCann became Chief of Staff to the Commander of the Tenth Fleet, reporting to Admiral Ernest King. The Tenth Fleet, known as "a fleet without a ship," was a highly specialized intelligence command able to summon a network of ships and planes (hunter-killer groups) for special assignments to thwart U-boat operations in the Atlantic.

The unit's success was augmented by information from deciphered German naval codes, allowing the strategic placement of intercepting ships to sink U-boats. Using hedgehog projectiles and depth charges, the group could neutralize one of the final threats to the East Coast and ultimately sank a number of die-hard U-boats.

As the war wound down in June 1945, Rear Admiral McCann was transferred back to the Atlantic Fleet and designated commander of a task force that escorted Harry Truman to the Potsdam Peace Conference. During the trip, he advised President Truman of the bombing of Hiroshima. For his efforts and past performance, McCann received a letter of commendation from President Truman and his second Legion of Merit.

After the trip to Potsdam in December 1945, Rear Admiral McCann was named Commander, Submarine Force, Pacific Fleet, his final wartime assignment.

During his Pacific and Atlantic submarine commands, McCann's undersea force sent hundreds of Japanese and German war and merchant vessels to the bottom of the sea. At the conclusion of war, he was involved in the destruction and scuttling of the Japanese submarine fleet.

Later, after the war's end, Rear Admiral McCann led a submarine squadron in the Bering Sea and successfully commanded the first submarine under the polar ice cap. He personally participated in the operation, helping the Navy better understand the effects of ice and cold on submarines. With the use of sonar and fathometers, the USS *Carp* could find surface areas to recharge its batteries and travel long distances under the ice.

For a brief period in 1948, McCann was assigned to the Navy's top planning body, the General Board of the Navy Department. Shortly afterward, the Admiral's reputation for his broad knowledge, impartiality, and attention to detail earned him the role of Navy Inspector-General responsible for operational and administrative inspections and investigations, specifically charges involving ethical issues.

One of his first charges was to investigate what has been called the "Admirals' Rebellion." The rebellion was a dispute between the Navy and Air Force involving the future use of carrier-based planes versus bombers. One side could see no need to build expensive carriers when Air Force bombers could deliver bombs. The controversy included unsubstantiated collusion charges and ultimately resulted in a shake-up of the Naval Command. Admiral McCann's objectivity helped to bring the brouhaha to a close. (The Navy won the discussion when the advantages of carrier-based planes become more evident during the Korean War.)

A March 1950 *Transcript* article documents the month that the Admiral was admitted to the Bethesda Naval Hospital for observation and treatment. It states he is bothered by a recurring ailment that developed during his wartime service in the Pacific. The article also notes that Admiral McCann has been in the Inspector General's position for nine months of a three-year assignment.

At fifty-four years old, Rear Admiral McCann decided to retire in 1950. He had given thirty-three years of service since graduating from the Naval Academy in 1917. Upon retirement, he was promoted to Vice Admiral in honor of his service.

After leaving the Navy, Admiral McCann and his wife Katheryne sold their home on Cherry Street in North Adams, initially moved to Florida, and then relocated to California to be near their daughters.

Vice Admiral McCann passed away at age 81 on February 22,

1978, at the Balboa Naval Hospital in San Diego. In honor of his and his family's wishes, Allan Rockwell McCann's cremated remains were to be buried at sea. In a fitting tribute to this naval warrior, it would be the modern USS *Sailfish* submarine, named after its predecessor USS *Sailfish* (formerly the *Squalus* that sank off the New Hampshire coast) that was to be tasked with the sea burial.

On March 13, 1978, with all-hands-on-deck, including an honor guard and bugler, the Admiral was accorded full military honors, and Allan's ashes were scattered in the ocean he knew so well. The Navy sent a letter to the family, with photos documenting the burial at sea.

For decades, the McCann Rescue Chamber brought hope of survival to submariners the world over. More modern-day versions still incorporate many of McCann's initial chamber features and can descend to even greater depths.

Technical Sergeant Michael F. Cirullo
Silver Star/Bronze Star
WORLD WAR II

He landed with Company D on Omaha Beach and, for the next nine months, led his heavy weapons platoon through four countries and hundreds of villages. They fought fanatical Nazi resistance, encountered frequent counterattacks, and endured near-constant artillery shelling. He earned the Silver Star twice, the Bronze Star, the Belgian fourregère from a grateful country, and several unit citations. More recently, in 2020, Michael Cirullo's division—the 30th, known as Old Hickory—was awarded the Presidential Unit Citation for "extraordinary heroism."

MICHAEL CIRULLO, A THIRTEEN-POUND baby, was born on February 12, 1913, most likely at home in North Adams to Francesco ("Frank") and Elizabeth ("Rosa") Cirullo.

Frank and Elizabeth were born in the late 1800s in separate small towns in the southern Italian province of Catanzaro, within the region of Calabria. Frank immigrated to the United States in 1898, attracted to North Adams by relatives and the outreach efforts of a local pastor. He returned to Italy some years later and met and married Elizabeth Lombardo, whom he called Rosa, in 1910. Soon after, he, Rosa, and their newborn daughter returned via Ellis Island to the growing North Adams community. Frank went on to serve in the United States military during World War I.

The couple became parents to seven more children, for a total of six girls and two boys, raising their lively family on a large plot of land on North Veazie Street that eventually became Myers Avenue. In addition to being a homemaker, Rosa tended her three vegetable gardens. Frank worked in the warehouse at the local Sprague Company.

The family grew much of what they ate. They had fruit trees and a grape arbor and also grew tomatoes, potatoes, string beans, and Italian squash. Rosa was an ardent canner of vegetables. The family also had a cow and a flock of chickens. Her son Michael was one of her hardest-working farm hands.

Michael would be educated in North Adams schools and attended the nearby Johnson School. He worked at the Hoosac Cotton Mills, Arnold Print Works, and Sprague Electric Company until the outbreak of World War II.

Records indicate he enlisted in the service at the age of 28 on August 31, 1942, had a brief furlough, then entered the US Army in its October draft quota and was inducted on October 31.

After basic training, he was assigned to Company D, 1st Battalion of the 117th Infantry Regiment. Mike was designated to join the heavy weapons platoon of his company and sent for further training to Camp Croft in Spartanburg, SC, before returning to his unit at Fort Benning, Georgia.

Mike back home on a brief furlough. *[Courtesy of Whitman/Cirullo family]*

In World War II, an infantry company had four platoons: three rifle platoons of approximately 40 men and a heavily armed weapons platoon. They were trained in using .30- and .50-caliber machine guns, 60- and 81-mm mortars, flamethrowers, and recoilless rifles (an anti-tank weapon.) The heavy weapons platoon aided in assaults or defense against the enemy.

The 1st Battalion spent several months in maneuvers in the hills of Tennessee near Camp Forrest, which was by that point serving as an internment camp for as many as 800 civilians of German, Italian, and Japanese descent, including American citizens. (In 1943, all civilians were transferred to other camps to make way for over 20,000 German prisoners of war.)

The battalion then moved to Camp Atterbury in Indiana for more months of tactical training before departing for their staging area in Massachusetts. The 1st Battalion arrived at Camp Myles Standish in Taunton, Massachusetts, in a cold, bleak January 1944 and embarked for England from Boston on the MS *John Ericsson* in mid-February. (MS *Ericsson* had been a Swedish passenger ship requisitioned by the US government to transport troops. It was returned to the passenger line after the war.)

After a twelve-day voyage on a crowded ship, they arrived in England in late February 1944 and settled in for many cold, damp months of training, punctuated by countless inspections. Soon after D-Day (June 6, 1944), the battalion was alerted and sent to Southampton, England, for embarkation to France. At that time, as described in *Curlew History* by William Lyman, each man was given 200 francs, a supply of cigarettes, and several boxes of K-rations.

The 1st Battalion had four Infantry Companies: A, B, C, D, and a Headquarters Company. Company C would be the first battalion unit to land on Omaha Beach on June 13. Companies A and B landed on the 15th, and Pfc. Cirullo followed with Company D on June 17, 1944.

An integral part of what was considered the beginning of the Normandy Campaign, Mike's unit began hiking inland, almost immediately passing churned-up landscape, flattened houses, destroyed vehicles, eviscerated, bloated German bodies, and dead cattle everywhere. As the smell of death increased, they knew the war drew closer.

Within days of landing, Cirullo became a sergeant; in his words, he "was promoted to buck sergeant because of the heavy losses of non-coms on the beaches." As his weapons platoon advanced, it actively supported the battalion's other companies, and Michael's Company D sustained its first casualties. Often, platoons or companies in supporting roles suffer fewer casualties, but this did not play out for Mike's company.

Hedgerows soon became their focus: These high mounds of dirt served as fences between plots of land topped with thick vegetation growth. They provided excellent defensive works for the German Army and were defended by German machine gun teams and tanks. In the early part of the battalion's advance, there was much "stop and start" because of the gnarly, wide, and high hedgerows.

Mike's heavy weapons team was in constant demand to put down suppressing machine gun fire or provide overhead mortar fire so groups of soldiers could overcome pockets of resistance. It was a dogged, relentless, murderous undertaking with no quarter given. Finally, working in tandem with tank units that used prongs welded on the front of their tanks, the hedgerow obstacles were destroyed, and resistance was obliterated.

As Company D fought on, they advanced in mid-July 1944 to a place known as Les Hauts-Vents, a crossroads in France referred to by the soldiers as Coffin's Corner. Mike's company came under heavy artillery fire and quickly suffered 34 casualties. Less than a month after landing, almost 20% of the company had been killed or wounded. Luckily, replacements arrived the next day. Mike was promoted to Technical Sergeant and assumed the responsibilities of a platoon sergeant.

They continued to advance through northern France, facing different degrees of resistance from retreating German troops. At one point, they could see the Eiffel Tower in the distance.

Company D reached Saint-Barthelemy on August 6; the Battle of Mortain began and lasted four days. It was a crucial battle of the war. The Germans hoped to cut off supplies of the advancing Allied armies using Panzer divisions supported by infantry regiments. Mike's 1st Battalion was in the center of the resistance; they held their position, and the heavy weapons platoon with their mortars and machine gun section played a large part in halting the onslaught and turning back the Nazis. Some mortar crews were said to be firing at tanks from a distance of 175 yards. The performance of Dog Company's machine gunners was noted as outstanding.

The 1st Battalion, for its intrepid spirit, was awarded a Distinguished Unit citation for its actions. The battle was not without cost; the battalion had over 250 casualties. Mike himself escaped unscathed and received a Silver Star for his valiant actions.

His citation read: "For gallantry in action from August 7, 1944, to August 11, 1944, in France. Sergeant Cirullo assumed command of his Platoon when his platoon leader and sergeant became casualties. When a general withdrawal of the Battalion was expedient, he so skillfully directed his men that he was able to furnish covering fire for the rifle companies at all times during this moment. In subsequent actions, Sergeant Cirullo directed mortar and artillery fire from hazardous positions, breaking up several enemy reorganizations, aiding in evacuating the wounded under severe fire. At critical points during this fighting, he maintained communications with adjacent units. Sergeant Cirullo's gallantry and heroism greatly aided in finally repulsing the enemy."

Company D continued at the forefront as the 1st Battalion surged through France and endured more losses than any other company. It spent the rest of August and most of September hiking, fighting, and liberating towns, moving across Belgium and then into Germany. The 1st Battalion became known by the radio codeword *Curlew*.

As the battalion approached Germany in late September, several more battalions were assembled to assault the vaunted Siegfried Line. The Curlew battalion was one of them. Mike's mortar and machine gun team would soon be in the thick of it, providing supporting fire while groups of soldiers tried to destroy the enormous concrete bunkers.

Company D soldiers established a dangerous process to destroy the bunkers. Under suppressing fire by machine gun, several soldiers would rush forward and throw grenades through bunker embrasures, then another soldier would jump atop the bunker and drop a grenade down the air ventilator. Next, the flamethrower team would squirt a burst of liquid into the embrasure while another soldier placed a satchel charge at the bunker's exit door. At this point, the remaining enemy would abandon the fortification, to be killed or captured.

As Company D assaulted the Siegfried line, Mike's heroism was again recognized by being awarded his first Silver Star.

Award of the Silver Star Medal

Citation

Technical Sergeant Michael F. Cirullo, 31128546, 117th Infantry Regiment, United States Army, is awarded the Silver Star for gallantry in action on October 2, 1944, in Germany. When his Company was pinned down by enemy fire during an assault upon a sector of the Siegfried line, Sergeant Cirullo risked his life to move his mortar section into a position from which they delivered deadly fire upon the enemy guns, enabling his Company to make a frontal assault and capture six enemy pillboxes. The courage, determination, and fighting spirit displayed by Sergeant Cirullo reflect great credit upon himself and the Armed Forces.

–L.S. Hobbs, Major General, US Army Commanding

Of note, Mike received a Silver Star Oak Leaf Cluster (representing a second silver star) for this action on August 7–11, 1944. It appears the paperwork for the August action was delayed and not discovered until after his October heroics. He received his first Silver Star award on October 2, 1944. The 1st Battalion, for its gallantry in action, received another Distinguished Unit Citation, and the 117th Infantry Regiment as a unit received the Croix de Guerre.

October 1944 found Company D continuing to advance

After being decorated for heroism, Mike is promoted. *[Courtesy of Whitman/Cirullo family]*

through towns and villages, routing dug-in German troops and resisting fanatical counterattacks while taking considerable casualties. They periodically paused for a day to receive replacements and then resumed at the same pace, always forward. The replacements were green and needed to learn quickly on the job if they were to survive.

On October 18, as the battalion approached the German town of Warden, Company D took casualties when it was repelled twice by heavy barrages from German self-propelled guns and tanks. The battalion counterattack cut off a company of fleeing Nazis. Mike's heavy weapons platoon was credited with killing sixty Germans, taking 145 prisoners, and capturing three tanks.

The company slogged on for another two months through villages and towns with little relief. They remained wary every step of the way, always wondering if even the small hamlets would have a sniper's nest or an armed Hitler Youth willing to sacrifice himself for the Third Reich.

On December 17, 1944, the men were looking forward to some rest, but it was not to be; the battalion was called into immediate action in response to what became the Battle of the Bulge, a last-ditch counteroffensive by Adolf Hitler to reverse his losing efforts in World War II. A tide of 200,000 German troops and thousands of tanks attacked Allied forces, seeking to drive them back to the French coast.

On December 18, Mike and his company found themselves on the outskirts of Stavelot, Belgium, staring at what looked like half the German Army; the streets were swarming with German soldiers and

Tiger tanks. Action was furious for the next several days as Company D played a critical role in fighting off six fanatical counterattacks by SS troops and Tiger tanks at the town's bridge area. Sergeant Mike Cirullo, his lieutenant, and several others were mentioned in dispatches as outstanding performers.

One day stood out, when Mike, his platoon leader, and several others advanced to a house filled with SS troops. The lieutenant called for a Sherman tank to fire several rounds into the first floor, then Mike and the lieutenant cautiously advanced, hearing some noise in the cellar. As they approached the house, the Germans fired up through the floor, and the lieutenant and Mike threw grenades down the stairs and called for their surrender. Several more grenades were thrown down the stairs for added encouragement, and twelve Germans plodded up the stairs with their hands raised.

While the Germans were climbing the stairs, from the second floor an SS trooper suddenly appeared with his machine pistol ready. Before he could shoot the lieutenant in the back, Mike leveled his carbine and snapped off two shots to the chest of the trooper. The German tumbled down the stairs, dead at their feet.

Searching the cellar yielded two dead Germans and one somewhat alive with his jaw blown away.

When marched outside under gunpoint, it was clear that one of the prisoners' arms had almost been severed by a grenade shard. One of the SS men asked for a knife, and with a swift motion, the arm was cut off and the battle wound dressed. When the Americans were distracted by the injury, one of the SS troops tried to escape and was quickly shot by one of the guards. The remaining 11 prisoners were marched off to captivity.

SS troops had occupied the town for days and were acknowledged to have executed over 100 civilians. Mike's lieutenant found the bodies. More than 20 women and children were killed in a garden several houses away from where Mike was fighting. During this same period, under the command of SS Officer Joachim Peiper, the same unit had committed additional war crimes in nearby Malmedy,

Belgium, when they machine-gunned more than eighty unarmed American soldiers.

As fighting continued, within a month's time Mike's platoon leader was wounded, and the group was now at half-strength due to casualties. Several sergeants (including Mike) took over and led the diminished platoon as it waited for replacements. Some headquarters personnel were moved in the interim to help bolster the 2nd Platoon's roster.

Mike was again recognized for his heroism and awarded a Bronze Star. His citation reads:

> *Technical Sergeant Michael F. Cirullo . . . is awarded the Bronze Star for heroic achievement in action on January 22, 1945, in Belgium. When Sergeant Cirullo was notified that his machine gun platoon was to be ready to give fire support during an attack, he reorganized his Platoon, as it had been depleted by casualties, and formed one section. After making a thorough reconnaissance for the most advantageous positions, he and his men prepared positions for four machine guns and at daybreak were ready to give supporting fire. All during the day, the enemy delivered heavy fire on the positions, but Sergeant Cirullo courageously kept his guns in position to render fire if needed.*
>
> *—L.S. Hobbs, Major General, US Army Commanding*

The unit had earned a brief rest period during February in Warden, Germany, then crossed the Roer River under heavy barrage of enemy artillery. The battalion resumed assaulting German positions and, for days, kept pushing into Germany, staying close to retreating German troops, always drawing and inflicting casualties.

Finally, after over nine months of almost continuous combat, Sergeant Cirullo rotated back to the United States on March 26, 1945.

A news article in his hometown newspaper states that

Mike standing in front of the Veterans Honor Roll, which listed local each servicemember's name on a small wooden slat. *[Courtesy of Whitman/Cirullo family]*

Sergeant Michael F. Cirullo, son of Mr. and Mrs. Frank Cirullo of 3 North Veazie St., has been honorably discharged from the service, with 88 points on the Army's point discharge plan. Sgt. Cirullo . . . holds the Silver Star, Oak Leaf Cluster to the Silver Star, Bronze Star, a unit citation, and the Presidential Unit citation.

After Sergeant Cirullo rotated back stateside, combat intensity began to wane and the war's end approached. The battle-hardened Company D had few remaining firefights, began capturing countless prisoners, and advanced miles through Germany daily. The company's final battle was at Magdeburg, Germany, on April 17 and 18, 1945. The war in Europe ended on May 8, 1945, and the division served on occupation duty until August. The division returned to the United States and was deactivated in November 1945.

All in all, Sergeant Michael Cirullo's unit was accorded the following honors:

- 1st Battalion was awarded Distinguished Unit Citations for its two roughest battles, those at Mortain and the Siegfried Line.
- The 117th Infantry was awarded the Croix de Guerre with a silver star for its action on the Siegfried line.
- 30th Division was cited for Belgian liberation and the Ardennes breakthrough and awarded the Belgian *fourragère* for both operations.

On March 17, 2020, the 30th Infantry Division (nicknamed "Old Hickory" after President Andrew Jackson) was awarded the Presidential Unit Citation for extraordinary heroism at the Battle of Mortain, France, in 1944.

COSTS: The 1st Battalion, while fighting in the five campaigns of Normandy, Northern France, Rhineland, Ardennes, and Central Europe, through the four European countries of France, Belgium, Holland, and Germany, liberated hundreds of towns and sustained more than 1600 casualties between June 1944 and May 1945. Company D suffered 204 casualties during combat, about 140% of its strength, with 27 men killed in action, 163 wounded in action, 13 prisoners of war, and one missing in action.

Mike was discharged in 1945, returned home, and found a job with a general contractor, Gordon and Sutton, in North Adams. He joined the local Laborers International Union and established a reputation as a hard-working, dependable employee.

He married Gina Irene Trenti at Saint Anthony's Church in 1951. Shortly after the wedding, Mike's dad passed away, and he and Gina decided to reside with his mother on Myers Avenue.

Elizabeth ("Rosa"), known to her grandchildren as "Grams," spoke Italian and understood only a smattering of English. She was renowned for her cooking. She made a special treat called by several names, most frequently *Fiori di Lucia.* Early in the morning Grams would pick orange female flowers from yellow squash and gently wash them. Then she would dip the flowers in a flour mixture, fry them in vegetable oil until they were light brown on both sides, and serve them in warm tasty patties that everyone loved.

Most weekends in the summertime, there were family gatherings at Grams' and at her daughter Rose's house across the street. Aunts, uncles, cousins, and all relatives were welcome. Adults wandered back and forth between the houses, men listening to ball games, women chatting, and kids running in the fields or woods. It seemed like there were picnics almost every weekend, but especially on Grams' birthday in August. The weekends were a great time.

Mike enjoyed working in Grams' garden and maintaining the property. The barn where he kept his tools seemed to be his place for solace, where he would smoke a Lucky Strike and spend some time alone. When not working around the house on Myers Avenue, he could be called into action to help one of his sisters with a project, building a wall or tiling behind a stove. Mike was handy and always willing to help.

He and Gina enjoyed going out to dinner and especially taking trips to the local racetracks for amusement. They would go for Sunday rides to Pownal, Vermont, to bet on the races or maybe wander over to Saratoga, New York, to wager on a favorite horse. They enjoyed being together and were usually in a Pontiac Bonneville. Mike liked big, easy-riding cars, maybe a natural preference after bouncing in the back of a stiff Army truck.

Mike and Gina retired in the late 1970s, he from Gordon and Sutton and she from the Sprague Electric Company. Mike passed away in 2001, Gina in 2014. Both rest today in Southview Cemetery in North Adams.

In his book *Curlew History,* William J. Lyman takes note of "Sergeant Cirullo, a first-rate fighter who had been in combat all the way through," an apt description of this young man who was in many ways like numerous others who served in World War II. He came home, got a job, married, didn't complain, seldom talked about his part in the war, and moved on with his life.

Mike never discussed his almost 280 straight days of combat. Once very late in life, at the request of a great-nephew who was recording World War II testimonies for a school project, he shared his medals and a swastika armband taken from a dead German soldier who'd almost killed him. Mike reminisced about walking through an area when a bullet zinged past his ear; his friend spotted the German before he could get off another shot and killed him, after which they relieved him of his armband.

What wasn't said then and can be said now is that this quiet, self-effacing man from North Adams earned two Silver Stars and a Bronze Star, was in almost continuous action over nine months, and in many cases, it was his heavy weapons platoon's efforts that helped to determine the course of a battle. He was a first-rate fighter by any measure, a survivor of a war that returned too few boys to their hometowns, and permitted too few men to live out their days in some peace.

Sergeant Michael Scarpitto
A Ritchie Boy
WORLD WAR II

Out of the 16,500,000 men and women who served in the US military during World War II, fewer than 20,000 graduated from Camp Ritchie, a newly formed secret intelligence group located outside of Washington, DC. Its graduates, dispersed among different military groups throughout the world, were assigned the formidable task of gathering intelligence that would help to win the war. A classified postwar Army report indicated that nearly 60% of the authentic intelligence information gathered in Europe came from the Ritchie Boys.

Michael Scarpitto, a single, thirty-three-year-old high school teacher from North Adams, was among this select, secret group of highly trained agents who performed heroically throughout World War II. Many were wounded, and still others were executed, giving their lives for their country.

MICHAEL'S LIFE BEGAN IN September 1909 on State Street in North Adams, Massachusetts. He was the youngest of five children born to immigrant parents, Antonio and Carolina (Tannacito) Scarpitto.

Both Antonio and Carolina were born in the same Italian village in the latter half of the nineteenth century. Family stories have it that Carolina, who was quite intelligent although not as literate, pledged to marry the most handsome man in the village and decided Antonio was that person.

In the late 1800s, the couple determined that Antonio should work in the United States and save money before returning to Italy. They were encouraged by recruiters looking for laborers for the Berkshire mills. After several years of work, Antonio decided he wanted the family to join him in the United States for a better life. Carolina, who already owned property in Italy, did not want to leave but acquiesced and joined Antonio in the early 1900s.

Over the next decade, the family grew to five children, and sadly Antonio's health deteriorated to the point that his labors were greatly restricted. In 1913, when Michael was four years old, his father died from tuberculosis, known then as consumption.

The family was now fatherless, with five children under ten years of age and a mom who could not speak, read, or write English.

Carolina, or "Nonna" as she was known to everyone, was a force unto herself. She worked odd jobs, cleaned houses, grew her vegetables, slaughtered her chickens, prepared all their meals, cut firewood, and even bought the two-story house they lived in on State Street. At times, she would bring Michael with her to work at the mill, but once he played with the sewing machine and got a needle through his finger, so the boss prohibited her from bringing him again.

Carolina Scarpitto remained self-sufficient her entire life. She eventually learned enough English to get by. She was never interested in applying for welfare (although she probably qualified) and wouldn't apply for citizenship until she could gracefully sign her name. To this day, the memory of Nonna garners respect and admiration from those who knew her.

Michael, a voracious reader from an early age, attended local schools and graduated from Drury High School in 1928. The yearbook reports that he played varsity baseball for three years and captained the team in his senior year. His position seemed to vary from shortstop to pitcher. It also notes, "His outdoor sports have not interfered with his scholastic standing, and Dr. Gadsby finds him a valuable asset in Latin."

Michael had the support and encouragement of several people, including a high school teacher who realized his academic capability. He took a year off to help support his family and then enrolled at Hamilton College. During the year off, Michael worked as a railroad laborer and mailman, giving his earnings to his mom.

Arriving at Hamilton with a scholarship and little personal money, Michael quickly got several jobs to support himself. He worked at a fraternity house cleaning tables and as a yard boy for a wealthy family. He loved the college environment but was lonely, having little money to be able to return home often. He majored in languages, played varsity baseball and basketball, became acknowledged as a Latin scholar, and was vice president of his senior class.

In a considerable turn of events, Delta Upsilon, the fraternity where he worked (but could not afford membership), got together and offered to pay his dues and initiation fees so he could join. He declined their generosity because he was unsure how he would pay them back. Later, Michael bought the library of the wealthy family for whom he used to do yard work, purchasing it on a payment plan.

Michael graduated in 1933 with an A.B. degree with honors, returned to North Adams, and taught for the next ten years in the public school system as a substitute teacher, teacher, and then assistant principal. He loved teaching Latin, French, and Italian and he also taught biology. While teaching, Michael completed the requirements for his Master of Education in Administration from Harvard, and he and his mother attended commencement exercises in June 1940.

When the United States declared war on Japan and Germany in December 1941, Michael was not drafted immediately. He continued to teach at Drury and live on State Street with his mother. Older, single at the time, and having taught for ten years, he was one of the few teachers with a car, although that would be sold after he joined the Army and went to Europe.

Michael enlisted in January 1943 and left for basic training at Fort Devens. For the next seven months, in addition to his rigorous introduction to the Army, he attended courses at City College of

New York on Testing, Classification, and Interviewing and at the University of Minnesota on International Relations and Languages.

It seems Michael had already been identified as a different type of soldier, at least ten years older than the typical teenage draftee, highly educated, and accomplished in four languages.

Before Michael's enlistment, but after the United States declared war, the military realized its intelligence services were woefully inadequate and needed a program to train its servicemen. Funds were authorized, and a secret intelligence training program was approved to be located at Camp Albert C. Ritchie in Maryland, named after the then-current governor of Maryland.

The Military Intelligence Training Center (MITC) at Camp Ritchie was opened on June 19, 1942. Those attending the camp eventually became known as "Ritchie Boys."

Located on over 600 acres, the camp had hundreds of structures and was, in effect, a mini-city with barracks that could house thousands of troops. It had accompanying mess halls, multiple training facilities, headquarters, and gymnasiums, as well as a post office, hospital, infirmary, theater, chapel, and post exchange.

During the war's most critical period, from 1942 until 1945, Camp Ritchie would graduate 31 specific classes, totaling over 11,000 soldiers, and train thousands more in various intelligence crafts. The eight-week course was rigorous; one-quarter of soldiers did not make it and returned to their units.

The intelligence camp established an ongoing nationwide search for anyone in the Army having foreign language fluencies such as German, Polish, Italian, and French. Those men would receive secret orders to Camp Ritchie, and Michael Scarpitto was one of the chosen.

Michael enjoyed the Camp Ritchie environment. The soldiers were often older, multilingual, intellectual, and included writers, artists, and teachers. The mess hall environment and conversation were lively and different from everyday barracks life.

PFC Michael Scarpitto was with 511 other intelligence students in Class 11. It began on August 25, 1943, and graduated 427 students on October 23, 1943. Seventy-four, including Michael, were classified as "Interrogators of Italian Prisoners of War" (IPW-It).

Mike as a young intelligence officer. *[Courtesy of Scarpitto family]*

Many of the Ritchie graduates became interrogators, although other courses were of great benefit to the Army. The students became experts at analyzing aerial photos and captured enemy documents. They came to understand what was called the German Army's "Order of Battle" and could advise commanders in the field of the enemy's size, capabilities, and weaknesses.

Michael, and many on his intelligence team, all of them linguists, were trained interrogators. They had spent many hours learning different techniques at Camp Ritchie. The most basic rule for interrogators was that a prisoner could not be physically harmed or humiliated. Tricks were acceptable, just don't touch.

They practiced interviewing each other in front of the class, receiving critiques, and the interrogations were also helpful just in case they were captured. The interrogators often got frustrated when trying to elicit information from their truculent friends. When food, coffee, and cigarettes didn't work, they might resort to the threat of sending them to Russia.

The Ritchie instructors acknowledged that trickery with prisoners was useful. Occasionally interrogators wore uniforms enhanced with ranks higher than those prisoners they were interviewing, a vital issue with German prisoners. Also, the threat of sending a German prisoner to the Russian front seemed to work. Over time, the Ritchie Boys became masters of trickery.

At graduation on October 23, 1943, all the graduates were warned not to tell anyone—even family members—that they would be working in Military Intelligence. In preparation for going overseas, Michael was sent to a four-week Combat Course. The course was sponsored by the Army's Counterintelligence Corps (CIC), which had been seeking highly intelligent recruits with foreign language skills in French and German. Michael, at this point, became a CIC operative.

Of note, among other investigative wartime duties, the CIC was chartered to prevent sabotage and espionage, capture enemy documents, and intercept spies. One of the more famous Camp Ritchie–trained CIC agents was writer J.D. Salinger, who not long after the war would publish *Catcher in the Rye.*

After training in marksmanship, map reading, patrolling, scouting, and enemy identification, Michael graduated in early 1944 and departed for England by boat on February 20, 1944. Thousands of American troops were massing for the upcoming invasion of the European continent.

Upon his arrival in late March, interrogators were assigned to six-man teams with a broad array of language skills. Michael's team was designated MI-514. Most interrogators had been sent to Britain for further training before the invasion of Europe. The British were uniformly recognized as having more developed military intelligence programs.

In their first months in England, the team spent long days interrogating prisoners of war at secret camps, sharpening their skills. Michael developed his style: friendly, showing genuine interest in the prisoners, and, although not a smoker, using cigarettes and especially chocolate, to encourage conversation. Michael also served as an interpreter between Italian prisoners and people assigning them work. The prisoners were paid in shillings that could then be sent home to their families in Italy, who were in dire need of support.

In England, Michael met a future team member who was to become a lifelong friend, Tore Mellgren from Scandinavia. They would be partners the entire time they were overseas. Tore first entered the United States by working on a cruise ship. He became a naturalized citizen, joined the Army, and was recruited for the Ritchie Boys due to his fluency in Swedish and Norwegian.

They were opposites. Michael tended to be quieter, liked to read, and was a negotiator. Tore, a little more flamboyant, loved to clean his pistol at night, hoping to be able to use it someday for the cause of freedom—and yet they would be close sidekicks for the rest of the war and long after.

After months in England, the team crossed the English Channel, entered France in the fall of 1944, and was initially assigned to interrogate POWs and French civilians as close to the front as possible, seeking information on German troop strength, morale, weapon placement, land mines—anything to aid troops and reduce casualties.

As the US Army pushed across Europe, the role of Michael's CIC team became investigatory. He was involved in conducting investigations of crimes, sabotage, sedition, and subversive activities involving military personnel or civilians. His team had the wide-ranging authority to detain or arrest suspects and was responsible for preparing case reports.

Michael was also involved in criminal investigations involving acts by American soldiers, although most often, he and others were trying to identify SS troops who had committed war crimes.

Their area of operations spanned France, Belgium, and Holland.

Michael himself traveled extensively in Belgium because of his fluency in French. His only means of transportation was a Willys Jeep that lacked a heater and once resulted in frostbitten toes.

Mike's Jeep, with his soon-to-be wife, Jacqueline Dubucs. *[Courtesy of Scarpitto family]*

After Liberation, the US government was asked to help Belgium reconstitute its government and police department, which were riddled with Nazi collaborators who had reported on Jews and resistance members during the occupation, most of whom were sent to their deaths. Michael was assigned the task of interviewing government employees and help to ferret out the collaborators.

Some studies indicate that the Belgian administrative system was very compliant and an instrument of collaboration during the war. In some cases, it encouraged citizens to join the German Army. It was a difficult situation, with the collaborators often identifying others to deflect blame from themselves. With his high degree of integrity, Michael did not want to surface names that were not true collaborators.

Michael knew the penalty for collaboration could be death, or at minimum, social ruination, so he was extremely cautious. He was determined to search for and find the truth. He hoped his IPW training and CIC work along with an engaging approach would be sufficient to determine who was lying. His IPW training had instructed him to look for tell-tale signs (a furtive look, sweaty brow, flared nostrils, a bobbing Adam's apple), but he knew there was more to it than just those indications.

Michael's caution was well founded: he had once witnessed a young Belgian man being executed by a firing squad for being a traitor. When Germans initially occupied Belgium, they had scooped up a slow-witted peasant (alongside many other young men) and given him a choice of working in a bomb manufacturing plant in Germany (that was itself being bombed frequently) or fighting the Russians on the Eastern front. This young man chose to fight the Russians, who eventually became an ally of Belgium, to his bad luck. When captured later in the war, he was labeled a traitor.

Michael never forgot the tragic scene of the boy being dragged before a firing squad, his wooden shoes falling off his feet while officials calmly watched his death.

Realizing his skills alone were not enough, he gained the confidence of two local, highly respected Belgians: a petite woman doctor who had worked with the Resistance and faithfully cared for the Belgian population during the war and a respected member of the local police force.

During wartime, the Germans had taken most of the male doctors to Germany, but females were left behind. She and her chauffeur, who was in the Resistance, had continued to make house calls. She cleverly had certified her chauffeur as mentally disabled, so while waiting for the doctor, he would listen to other German drivers and gather intelligence for her.

The doctor would also treat wounded Resistance fighters and downed Allied airmen in the evening. She was eventually honored at war's end by both Belgium and Great Britain.

As Michael's vetting process continued, he would consult with the esteemed doctor. Even though she hated what the Germans and collaborators had done to her country, she would not turn over their names to Michael. Still, if Michael provided a name, she would confirm an accurate suspicion with a nod.

Amid this vetting of collaborators, on one rainy, overcast day mid-December 1944, Michael was able to visit the Notre Dame Cathedral in Paris, only to find it locked. In French, he asked a nearby pretty mademoiselle if she knew why the church was closed; she replied in slightly accented English that she did not, and so began a twenty-minute conversation that led to a fifty-five-year love affair. Her name was Jacqueline Dubucs.

The happy young couple on the streets of Paris. *[Courtesy of Scarpitto family]*

After their conversation, Michael returned to the barracks and told Tore he had met the most beautiful woman in the world, and he was going to marry her, although he did not know her address. Eventually, Michael found her, and they met whenever possible between his assignments, almost always with Tore as his too-faithful chaperone.

At times it felt awkward for Jacqueline to have two handsome soldiers, one on each arm, strolling down the avenues. It could feel uncomfortable, like she was fraternizing with American troops. She returned temporarily to the city of Bayonne, located in Vichy, France, and continued to work in her family's import business. It wasn't long before her dad passed away from pneumonia.

Michael continued to work primarily in Belgium, seeing Jacqueline whenever the opportunity arose. Often, she would need to travel to Paris to meet, since Bayonne was still under a dubious government.

After months of dating (and presumedly Jacqueline's in-depth vetting process by the CIC), the young couple was married at the place they first met, Notre Dame. It was June 16, 1945. Their wedding picture shows a handsome, smiling, uniformed soldier with a beautiful, fashionably dressed lady wearing a white suit and hat. The suit sports a black-trimmed collar in memory of her recently deceased father.

On the steps of Notre Dame, the beginning of a storybook marriage. *[Courtesy of Scarpitto family]*

The announcement in the *Transcript* stated they would make their home in Charleroi, Belgium, pending Sergeant Scarpitto's return to the United States. Much had occurred in Michael's life since he had left 183 State Street in North Adams as a single teacher 30 months earlier.

In December 1945, Michael departed Europe, heading home without Jacqueline; war brides would come later on a separate ship. Michael was honorably discharged in late December at Fort Devens' Separation Center.

His service earned him the Good Conduct Medal, an American Theater Campaign Ribbon, a Victory Medal, and the European African Middle Eastern Theater Campaign Ribbon.

After VE Day, many Ritchie Boys continued to work in Europe, interrogating civilians, working with governments, uncovering collaborators, determining who was eligible for state positions, and continuing to search for war criminals. Some detained SS war criminals or sent them to "denazification" camps.

Michael returned to live at home on State Street, awaiting the arrival of Jacqueline, and resumed teaching at Drury High School, coaching part-time. Jacqueline arrived on a boat reserved for war brides in April 1946.

After her arrival, the Scarpitto family grew; three girls and a boy were born over the next four years. In such a small, close-knit town, the hospital would call the school whenever the children were born, to let Mr. Scarpitto know of the birth. And his response would be that he would be there right after class. Everyone knew there would be no homework that night.

The family of six now shared the downstairs of the two-story home with Nonna. She continued to do all the cooking. The kitchen was Nonna's sanctuary. It was different for her, having another woman in the house, especially a foreigner who did not share her Italian ancestry. Still, they came to understand each other, and Nonna could see Jacqueline was good for her son.

In the early 1950s, Michael's love of education and desire for more responsibilities prompted him to begin studying for his doctoral degree. He was named Principal of Houghton School, although not long afterward he resigned his position, and the family moved to Chicago for his doctoral studies. They returned to State Street in the summers so Michael could work.

As his studies progressed, Michael was offered and accepted the position of Assistant Superintendent of Public Schools for Joliet, Illinois, and while there he helped them win the coveted All-American Award. While at Joliet, he completed his doctoral studies and, in 1955, received his Ph.D. in School Administration from the University of Chicago.

Over the years, the family always found time for summer vacations in their yellow-finned Shasta camping van. With somewhat tight quarters, the family of six plus a dog and a cat would find themselves traveling cross-country to Yellowstone National Park, or Canada, or once throughout the Southern states, books always at the ready.

All four of Jacqueline and Michael's children, Caroline, Renee, Michlyn, and Michael Jr., became avid readers and eventually language majors in college, just like their dad.

During this time, Michael's reputation as an educator garnered him the position of Superintendent of Schools in Stoneham, Massachusetts, a position he would hold for the next thirteen years until his retirement in 1970.

His influence in Stoneham was felt quickly. Within his first year, *The Boston Globe* was calling him a "fireball for education" for introducing a number of new admirable and impactful programs. The newspaper labeled it the *Scarpitto Revolution*.

His "Reading for Pleasure Program" encouraged first-graders through high schoolers to read any book they were interested in, urging teachers and parents to get involved. Dr. Scarpitto would visit classrooms monthly to celebrate ambitious readers. Teachers posted decorative wall charts tallying the number of books read. Parents helped with their children's reading assignments at home in the evening.

Michael's reading challenge became widely acclaimed and emulated and was noted in a *Time* magazine article. The method, which required no book reports, drove up book circulation at the library by 50% and thousands of additional books were checked out year-round; measurable comprehension went up, as did grades overall, not only in reading scores (some of the highest in the state) but in history and mathematics.

The results reaffirmed a comment he made to the *Globe*: "My job as I see it is to develop intellectual curiosity, a respect for knowledge, and academic achievement in these youngsters. And the more they read, the better their schoolwork will be."

Dr. Scarpitto also had Stoneham High School introduce a varsity "S" letter for all A students and members of the Honor Society. He was quoted as saying that "every child who comes through the Stoneham schools should get an intellectual workout."

To improve teacher recruitment, he led a program that focused on liberal arts colleges, identifying graduates with strong academic backgrounds. Next, he introduced a French after-school program for third and fourth graders and initiated Stoneham's first tuition-paid summer school program. Always a strong athletic booster, Michael helped organize football parades, sponsored a father-son flag football program, and upgraded playgrounds.

After thirteen years of serving the Stoneham community, the former Ritchie Boy retired. However, he stayed involved in education by consulting, substitute teaching, or serving as an interim principal. He and Jacqueline traveled throughout the United States, Canada, Europe, Mexico, and Great Britain.

In 1998, near the end of his life, his alma mater, Hamilton College, awarded Dr. Michael Scarpitto the prestigious Alumni Medal for devoting his life to childhood education. Michael passed away in 1999 and was returned home to North Adams to be buried in Southview Cemetery.

In 2000, Stoneham High School began and continues to this day to honor Michael with the annual Dr. Michael Scarpitto Scholarship, awarded to a highly successful academic senior. His son Michael Jr. has the honor each year of making the $1,000 presentation.

Jacqueline, his beautiful wartime bride, passed away in 2003.

Michael Scarpitto's story is one of warmth, love, wartime intrigue, and adventure. If it were in a library, it would be checked out again and again: A boy from a small town is recruited into a secret intelligence agency, shipped overseas into the center of a world war, and meets a beautiful French woman. He experiences love at first sight, marries, makes it home safely, raises a family, and becomes a beloved educator in his own home state, passing on his thirst for knowledge to a new generation of adventurers. It would be a blockbuster hit.

Lt. Colonel Ruben W. Shay
Guerilla/Special Forces
World War II, Korea, and Vietnam

[Manila, Philippines, Dec. 8, 1941] My father woke me up before 6 a.m. local time with the words, "The Japanese bombed Pearl Harbor last night—we'll be next. You'd better get dressed and report to your duty station." I was 17, one month short of graduation from high school. I was also an Eagle Scout and a member of the Senior Scouts Emergency Service Program.

My emergency duty station was the Red Cross Headquarters, just a few blocks up the street from my school. It seemed unreal! I ate breakfast, gathered up my books, and went to school. And there, reality set in. All of us senior scouts were told by the principal to report to the Red Cross immediately.

By the time the Japanese made their first bombing runs that fateful morning, we had been organized into teams, each with a station wagon, a driver, and a nurse. As soon as the first wave had passed, we were deployed to the areas of devastation with the mission of picking up those casualties still alive, rendering first aid, and [getting] them to the nearest emergency hospital.

—As told by Ruben Shay to *The North Adams Transcript*, 8-18-1995

RUBEN SHAY'S STORY BEGINS much earlier, in Germany. His grandfather, Ludwig, who spelled the family name *Schay*, was a prominent industrialist who owned shoe factories in Strasbourg and

Metz and later moved to Cologne, where he owned several apartment houses.

Ruben's father, Rudolf Schay, was born in Metz in 1899 and was wounded during World War I. After the war, Rudolf moved to Cologne and attended the University of Cologne, receiving a Ph. D. in 1921. He met and married Else Jakob in Cologne, where Ruben was born on February 24, 1924.

Rudolf, an accomplished young journalist, was purported to have interviewed Adolf Hitler in his jail cell after the failed Munich Putsch in 1923. In 1927, the family moved to Berlin, and as a political editor, Rudolf began writing critical articles about Hitler and the Nazi party. Conditions worsened as civil liberties became more restricted under the Nazi regime.

In 1933, around the time of the Reichstag fire, when the German Parliament was set ablaze and cited as cause for greater suspension of civil liberties, nine-year-old Ruben and his family quickly emigrated to France. Work was scarce in France, so they decided to join relatives in the Philippines. Both sides of Ruben's family were Jewish and felt the rising danger in Europe, but Ruben's maternal grandparents never made it out of Germany and perished in a concentration camp.

Ruben arrived in the Philippines at eleven. It was an American Protectorate at the time, and he attended American public schools administered by the US government, quickly learning to speak, read, and write English in addition to German and French. Ruben also became fluent in Tagalog, one of the two primary Filipino languages. Due to his travel, he was late to enroll in classes and chose Home Economics over study hall, where he learned the basics of cooking and baking, which he would appreciate later in life.

He joined the Boy Scouts, became an Eagle Scout, and subsequently entered high school. As part of the Boy Scouts, he became a senior patrol leader, a member of the Emergency Service Patrol, and underwent first-aid training with his cadre. Following his father's footsteps, he also became the school's newspaper editor.

Ruben's high school activities were dramatically interrupted when the Japanese empire attacked the Philippines on December 8, 1941. Emergency Patrol was immediately activated, and seventeen-year-old Ruben was pressed into service. With little sleep for the next three weeks, his team retrieved battered, bloodied, and broken bodies from the streets of Manila and transported the patients to the nearest hospital.

Often the station wagon turned into a hearse, with little that could be done for their patients. When Manila was declared an "open city" on December 26, 1941, the group was dismissed, and they turned in their gear and station wagon as the Japanese Army approached.

The "open city" declaration acted as a prelude to surrendering, intended to cease hostilities and protect public areas and civilians from further harm. It did not work in Manila. The Japanese continued to bomb the city.

Ruben then took the opportunity to join the nearby 31st Infantry Regiment and was assigned to Company G as a medic. At the same time, General Douglas MacArthur ordered the 31st Infantry Regiment to cover the hasty retreat of all military units in the Philippines to the Bataan peninsula. The group was loaded on a truck and headed for Bataan.

The delaying action worked for almost four months before the 31st was overcome by overwhelming force, lack of supplies, poor weapons, and rampant disease. The regiment ultimately surrendered in early May 1941. By the time the war ended, over 1000 soldiers of the 31st Infantry Regiment would have died in captivity.

While on the truck retreating toward Bataan, a major from Ruben's unit, seeing the hopelessness of their cause, ordered Ruben and two of his friends to leave the company in order to "evade the Japanese and join the guerillas."

Changing out of his uniform to peasant clothes, Ruben made his way back to Manila and contacted friends who put him in touch with the guerilla organization. Speaking Tagalog was a decided advantage for Ruben. He never saw his two friends again.

In his unpublished memoirs, Ruben notes that, upon his return to Manila, he was put in touch with a "guerilla band led by 'Miss U,'" whom he never met, and in the future he took all his instructions from a Filipino "Capitan." Ruben worked with the guerillas for the rest of the war, until the Americans liberated the Philippines in July 1945.

During those years, Ruben's band of guerillas participated in ambushing small Japanese patrols. However, the group also had him work in a local bar frequented by Japanese naval and merchant marine officers. He would use his limited Japanese to listen to their conversations. Then he would forward what he had overheard about their shipping plans to his group, and presumably, they would send it via radio to US Naval Intelligence.

His memoirs also recall a time when he was ordered to deliver a message to prisoners at the infamous Cabanatuan prison, where thousands of American Bataan death march survivors were being held. He accomplished the dangerous task of traveling over a hundred miles north of Manila with a Filipino friend, successfully avoiding roaming Japanese patrols.

Ruben's memoirs do say that one time he and a biracial traveling companion were apprehended and searched by the Japanese. He does not know what happened to his friend, but Ruben was jailed in the notorious Bilibid prison and placed in solitary confinement. The interrogator knew Ruben's father before the war, a stroke of luck that Ruben believed eventually led to his release.

In a *Transcript* article from December 2005, Ruben recalled that his main function was as a "corridor messenger": "It was dangerous, and I saw a lot of casualties in and around Manila. . . . I grew up very fast back then."

The month-long battle in early 1945 between the Americans and Filipinos against the Japanese to recapture Manila resulted in the deaths of over 100,000 civilians and destroyed most of the city's architectural and cultural landscape. It had formerly been known for its beauty as the "Pearl of the Orient." During this bombing and artillery slugfest, Ruben's house was burnt to the ground, and he was wounded.

When US forces finally entered Manila, Ruben guided some of the US Army's first patrols and was then assigned to the 29th Evacuation Hospital as a medical technician. At several different hospitals where battles were still being fought, he learned to assist in surgical procedures.

While Ruben had been fighting as a guerilla, his dad Rudolf served covertly as an agent for the US Army Counter Intelligence Corps (CIC). Once Manila had been returned to friendly forces, and during the time that Ruben was with the 29th Evac, Rudolf passed away. He was buried locally.

As normalcy began to return, Ruben was appointed city editor of the *Daily Pacifican*, a military newspaper circulated to every unit in the Pacific Theater.

At twenty-two, Ruben was sponsored by several people, left the Philippines, immigrated to America, and became a naturalized citizen. At some point during this time period, either when Ruben left Manila or when he completed his citizenship papers, he changed his name from "Schay" to "Shay." He has told some people that "no one knows how to pronounce *Schay* correctly," or he may have wished to move away from any ethnic discrimination.

During this brief trip to the United States, Ruben married a former classmate, rejoined the Army, and was sent to Japan for Occupation Duty. While in Japan, Ruben, who spoke fluent German, was used to repatriate German nationals who were living in Japan.

Upon his release from military service, Ruben returned to the United States. He attended the University of Michigan, majoring in history. He supported himself and his family by working odd jobs as a hospital orderly, gas station attendant, and bellman.

The military was never far from his mind, and with the US fully engaged in the Korean War, Ruben reenlisted in the US Army in 1951. Initially, he worked as a corporal in the Allied Liaison section because of his knowledge of European languages. With his talent, he was quickly recommended for Officer Candidate School and attended and graduated as an Infantry Second Lieutenant in June 1953.

Shortly after graduation, Ruben completed Parachute School and Ranger training and was assigned to the 505th Regiment of the 82nd Airborne Division at Fort Bragg, North Carolina. He served as a platoon leader and then executive officer of Company A.

Ruben completes Ranger training. *[Courtesy of Shay family]*

In the fall of 1954, Ruben joined the Army's Special Forces. His training was extensive; beyond attending jungle survival school, there was also demanding field indoctrination in special warfare techniques, advisor responsibilities, and continuous, stressful physical training. Individuals needed to be in superb physical condition, never knowing where or what climate they would be called to serve in, without notice.

Once his initial training was completed, he was assigned to an A-Team composed of twelve individuals consisting of two officers and ten enlisted men. Each person had at least two specialties in the use of weapons, explosives, communications, or medical assistance.

As recounted in his memoirs, one time his A-Team of twelve soldiers was training at Camp Lejeune, the Marine Base, when their Master Sergeant forced-marched the team over 120 miles back to Fort Bragg for what he considered a "short" conditioning march. Ruben remembers a restaurant along the way on one of the back roads refusing to serve the team's African American member. In solidarity, the team left the restaurant and ate their rations along the road.

Always training, Ruben was sent to the Army's Winter Mountaineering School at Camp Hale, Colorado, where he learned to ski, snowshoe, rappel in the snow, and master the art of winter camouflaging and other survival techniques, all while sporting huge packs.

It is notable that the Special Forces Group can trace its origins to behind-the-lines activities carried on many years ago by groups like the Philippine guerillas of whom Ruben was a member. It now has a worldwide reputation for conducting unconventional warfare and has operated and assisted many countries worldwide.

The 77th Special Forces Group (Airborne) to which Ruben belonged was activated at Fort Bragg, North Carolina, in November 1953. In 1960, it was reorganized and redesignated as the 7th Special Forces Group (Airborne), 1st Special Forces (also known as the 7th Group).

While training at Fort Bragg, on a one-night parachute jump, Ruben became disoriented and missed his designated landing zone, landing hard on an asphalt road. He broke his collarbone and several ribs, but he still assembled his team and completed the training exercise in record time.

After attending Pathfinder School and an Infantry Officers Communication Course, Ruben was sent to Korea in 1956 and assigned to be Company G Commander in the 31st Infantry Regiment, his old unit from the Philippines. Ruben spent the next two years upgrading the company's poor performance. Before Ruben's return to the United States, Company G received regimental and divisional honors for its high performance.

Over the next three years, until his next overseas assignment in Laos in 1961, Ruben found himself attending several Army schools, serving as an instructor at Fort Benning, and attending the University of Omaha, earning a bachelor's degree. His memoirs note that he was able to converse with his professors in German, French, and Spanish.

Promoted to Captain, Ruben returned to Special Forces at Fort Bragg to teach and train and then received orders to Okinawa in the spring of 1961. Only there for a few months, his A-Team was sent to Laos as part of Operation White Star, a classified operation (called a *military advisory mission*) during the earliest years of what would become the Vietnam War. The team's purpose would be to train the Laotian Army and local villagers to fight against a communist insurgent group known as the Pathet Lao.

The communists increasingly used Laos as a staging and transit area to resupply operations in South Vietnam. Ruben's team was helicoptered in to a remote Laotian village and, over the next few months and with the village chiefs' help, built a school and an aid station and trained local people in the art of self-defense.

Ruben commented in his memoirs that while living in the village, most of the team grew mustaches, something not usually allowed back on base. The natives began calling him *Capitaine Moustache*, a title he enjoyed but relinquished when his unit eventually returned to Okinawa and was ordered to "clean up."

White Star officially ended in 1962, and Ruben's team began working with the Laotian Parachute Regiment, conducting classes on battlefield tactics and then going to the field and using the techniques they had just learned in combat against the Pathet Lao.

The A-Team and their "trainees" often conducted combat airborne operations that quickly devolved into full-blown firefights requiring resupply by air, medevac, and often extraction under fire.

One of the group's main missions was reconnaissance to

determine Pathet Lao movements, uncovering and locating their supply trails. Once discovered and mapped, Ruben would travel to Thailand and report his unit's findings to the CIA.

After his experiences in Laos, Ruben, recently promoted to major, traveled to different units throughout Korea, Japan, and Hawaii, sharing his unit's guerilla warfare experiences with other combat units in the Far East.

When the lecture tour was completed, he was sent to Vietnam as an intelligence officer. Ruben's primary responsibility was to analyze intelligence reports. Never one to remain in an office for long, he decided to get his information firsthand. Usually, with only a pilot in an unarmed two-seater Cessna, known as a Bird Dog, he would fly low over enemy territory trying to determine enemy positions, often drawing enemy fire. During one of his excursions, flying at tree top level, they located the enemy when Ruben realized bullet holes were being punched into their thin-skinned aircraft.

Ruben Shay, left, preparing for a hazardous reconnaissance mission, Vietnam 1963. *[Courtesy of Shay family]*

Quickly breaking contact, they scrambled to a nearby base, had repairs made, and returned to the site to get the exact coordinates for a follow-up airstrike. For their bravery, Ruben and the pilot received air medals with "V" for Valor for their actions.

At the end of this overseas tour, Ruben attended the prestigious Command and General Staff College at Fort Leavenworth, and his family joined him on base. After graduation, his assignments for the next several years involved attending Army schools and instructing other units on unconventional/special warfare, his area of expertise.

In 1966, he was ordered to Vietnam, and because of his experience and rank, he was tasked with upgrading a Special Forces "C Team" detachment near Saigon.

Once that assignment was successfully completed, Ruben began working for Colonel John Singlaub (later to become Major General), a renowned, unconventional organizer of covert operations along the Ho Chi Minh Trail and neighboring Laos.

Colonel Singlaub assigned Ruben as the Liaison Officer to the South Vietnamese Special Forces Commander, a challenging, high-stakes job supporting hundreds of commandos working behind enemy lines. Ruben worked with Vietnamese and American commandos under the obscurely titled *Studies and Observations Group* (SOG) from 1966 to 1968.

The Studies and Observations Group (SOG) was a highly classified unit that operated in the Vietnam War. Its small groups conducted covert, across-border operations into neighboring countries of South Vietnam, including Laos, Cambodia, and North Vietnam. With the help of airpower and artillery, the group destroyed enemy supply lines, inflicting enormous casualties and disrupting communications. These groups of often unsung heroes were members of one of the most decorated units to serve in Vietnam.

The SOG was purported to have the highest kill ratio in Vietnam. For every 100 enemies killed, they lost one of their commandos. The high enemy casualty rate was at a serious cost to themselves; some commando units earned Purple Hearts over 100% of their headcount.

The group's willingness and daring to operate behind enemy lines far from direct support, often under intense fire, caused them hundreds of losses over the war years and resulted in a good number of Missing-in-Action, still unrecovered as of today.

Ruben continued to be heavily involved in the planning and support of inserting small groups (often fewer than ten people) of heavily armed men by helicopter, miles behind enemy lines in the most perilous situations. Once teams were inserted in enemy territory, they were frequently pursued and often on the run from much larger forces. Often in almost hopeless situations, they would run to a hastily created landing zone and shoot their way out to fight another day, taking their casualties with them.

Ruben, back from his last Vietnam tour. *[Courtesy of Shay family]*

After spending thirty-six months in Indochina, Ruben ended his tours of duty with two Bronze Stars, a Joint Service Commendation Medal, an Army Commendation Medal, and a Republic of Vietnam Gallantry Cross with Palm. His other decorations/awards include the Purple Heart, Combat Infantryman's Badge, Master Parachute Wings, and numerous campaign ribbons.

When Ruben returned to the United States in 1968 as a Lieutenant Colonel, he served as a senior advisor for a National Guard Battalion in Massachusetts and Connecticut. He assisted in their field training and tactical exercises and was awarded the Meritorious Service Medal.

After twenty-five years of arduous service, most of it with the vaunted Special Forces, Ruben honorably retired on May 31, 1972.

After retirement, Ruben and his wife Barbara moved from Connecticut to the town of Florida, Massachusetts, and purchased forty acres of land. The vaunted Special Forces career soldier set out to become a farmer. He and his wife, Barbara, created "Bear Meadow Farm," improving the land, clearing fields, creating gardens, planting thousands of herbs and flowers, and growing vegetables.

Barbara also made award-winning jams, jellies, and baked goods, and the couple produced maple sugar at one point. It wasn't long before the couple had a retail and mail distribution business selling herbs and preserves.

Over time, they added livestock and goats, chickens, geese, ducks, and honeybees to their farm. Eventually, after eleven years, as they grew older and the successful business grew more demanding, they sold the farm and moved to North Adams.

The move to town away from their numerous farm chores and off Florida Mountain allowed them to become more involved in a number of organizations. They quickly began renovating a large Victorian house on Wall Street, in the process updating and enlarging the kitchen, meeting the needs of Ruben's cooking and baking hobby.

Ruben, in full dress uniform, with his many medals and honors. *[Courtesy of Shay family]*

Ruben and Barbara both became very active in veterans organizations. At different times, each held leadership roles in the American Legion at the post and district levels and served on many boards. Ruben was an honorary and life member of the American Legion and served as the North Adams Post Adjutant, then Post and District Commander.

The American Legion Awarded him the "Outstanding Naturalized Citizen" plaque and the Conway Trophy, the highest award bestowed by the

Department of Massachusetts for Community Service. He was often called up to give Memorial Day and Veterans Day addresses and was pleased to wear his uniform.

Never one to remain still for long, Ruben and Barbara traveled in their Winnebago, took up boating, played golf, and snow skied. In their later years, Barbara became ill, and after a lingering illness, she passed away right before their 51st anniversary.

Shortly after Barbara's death, Ruben sold the large Victorian home, downsized to a smaller house with a large kitchen to accommodate his culinary efforts and continued to bake his bread, cookies, and cakes for family members.

He remained very active in the Legion. He enjoyed being involved with the Clark Art Institute in Williamstown, MASS MoCA in North Adams, The North Adams Public Library, and The Metropolitan Opera Guild.

Late in life, Ruben met another octogenarian, Joan Simpson Burns from Williamstown, Massachusetts, introduced by a mutual friend. It was perhaps an unusual match: a decorated Special Forces soldier who was a Jewish, German refugee from the Holocaust and a highly sophisticated New England socialite. Yet they enjoyed each other's companionship, fell in love, and were married.

It wasn't long before they expanded Joan's kitchen, and the smell of homemade bread wafted through her house. Ruben's love of history and biography matched well with Joan's editorial background, and they always had much to talk about.

They remained together for over ten years until Ruben passed away in 2017. He was buried by his Legionnaires in Eastlawn Cemetery in Williamstown, Massachusetts, with full military honors.

Although he did not talk often about his World War II or Vietnam experiences, it is no surprise that Ruben was described in the words of a close family member as "intensely patriotic, loved his new country, and proud of his military service."

In 2008, Ruben had written a memoir, *A Charmed Life*, which remained unpublished but provided a valuable narrative from which

Ruben's final resting place. *[Mike Remillard Photos]*

this chapter was drawn. One relative noted that the title *A Charmed Life* would have one think that his life was without challenges, perhaps a gifted life. Yet from Ruben's perspective, he may be saying how lucky he was to be alive. From his urgent, accelerated experience as a teenager, fighting over three years with Filipino guerillas in World War II, to his time later in life, serving another three years with Special Forces in Indochina, his temperament, skills, and wisdom were a priceless asset to his compatriots, and to his adopted country. It is we who are charmed by his life.

Corporal Gordon A. Greene
25 Infantry Division
World War II & Korea

"City's First Korea Martyr Home Today" read the headline of The North Adams Transcript *on Tuesday, August 28, 1951. Corporal Gordon A. Greene, a decorated soldier with the 24th Infantry Regiment of the Eighth Army, was the first of twelve North Adams men to die in the Korean War.*

Gordon died after almost nine years of military service, serving in segregated units the entirety of that time. He had also fought heroically in World War II's bloody African and Italian campaigns as a member of the 92nd Infantry Division, the only African American division to see combat in the war.

ON MARCH 23, 1924, Gordon Anthony Greene was born at home to Anthony Greene and Vivian Vanderburgh Greene in North Adams, Massachusetts, his middle name passed down from his father.

Anthony Greene had been born in Washington, DC, in 1900 and moved to Williamstown as a child. He graduated from Williamstown High School, where he was a star athlete and played varsity football. Anthony worked for the Massachusetts State Highway Department for over forty years, receiving a gold watch when he retired in 1965. He lived nearly twenty more years, passing on in 1982.

Vivian Vanderburgh (Lewis) Greene, had been born in Bennington, Vermont, in 1889 and married Anthony in 1919. For over fifty years, the couple resided in North Adams and ultimately had seven children, two sons and five daughters. After Gordon's death, Vivian became a Gold Star mother and was a member of the Veterans of Foreign Wars Auxiliary.

Little is known about Gordon's paternal grandparents, Charles and Margaret (Henry) Greene, who were both born in the Virginias or Carolinas in the 1800s then moved to and resided in Washington, DC. Gordon's maternal grandparents were Walter J. Lewis and Elizabeth Lewis Vanderburgh-Dawkins.

Vivian, Gordon's mother, was raised by Elizabeth and her second husband, Thomas Dawkins, who married in 1908. Thomas Dawkins had been born in Union, South Carolina, and worked for forty-three years for the Pullman Company as a porter on trains of the Harlem division of the Boston & Albany railroad. His route was between New York and North Adams, and the round-trips attracted him to the North Adams area, although he had little time to enjoy its beauty, with traveling so often. He retired in July 1948 and passed away a decade later.

Thomas and Elizabeth, Gordon's grandparents, were among the first African American property owners in North Adams. The couple lived on Walnut and Houghton Streets for most of their lives, Elizabeth passing away in 1966 at their home on Houghton Street. Elizabeth had managed their apartments on River Street, and in the summertime, she had rented rooms at their Walnut Street property to wealthy black New Yorkers.

Gordon grew up at 12 Washington Avenue with his parents and six siblings. The yellow, two-story slate-roofed house just off Ashland Street was said to be in "the swamp" because it frequently flooded, as it was located at a low spot between Church and Ashland Streets.

He attended Mark Hopkins Elementary School, was active in its recitals, and had the role of a waiter in the school's sixth-grade production of *The Frog Prince*. Several times he was commended for perfect attendance.

At home, chores were clearly defined: Gordon's five sisters cleaned the kitchen, washed the stairs, and did general housekeeping. Gordon and his brother Quentin managed a *Transcript* newspaper

paper route year-round and shoveled snow in the wintertime to earn extra money. (Any extra money earned by the children was brought home and given to Mom.)

Vivian was the family's cook and a good one. Breakfast was cereal or oatmeal. Since Anthony worked varied hours for the highway department and was usually gone for supper, mom and the kids ate spaghetti and meatballs, meatloaf, and different macaroni dishes. The best dessert was Mom's doughnuts made from scratch. The doughnut holes were reserved and used as a reward for children doing their chores.

The family dutifully attended St. John's Episcopal Church services every Sunday. Vivian ensured family attendance and that everyone was wearing their best clothes, especially for Christmas and Easter services.

Gordon had several nicknames: his family often called him Gordie, and others called him GAG (pronounced *gaug*) using his initials. Both were terms of endearment and respect for a young man with a low-key personality who was friendly toward everyone. With his dad gone a lot, he assumed many household responsibilities at an early age.

The family recalls a time when Gordie's mom sent him to the basement to check on the fuel level in the kerosene tank. As the oldest son at twelve, it was his responsibility. Since the basement did not have any electricity, Gordie lit a stick match to check on the fuel level, and there was an immediate explosion, causing the kitchen floor to leap up about four inches. His mom, the family's primary disciplinarian, paddled him soundly when he got back upstairs, even though his eyebrows and the hair on his arms were singed from the fireball.

Gordie and his sisters were active members of Sigma Lambda, a local social/charitable club that sponsored guest speakers, suppers, dances, and hayrides. The club disbanded in 1942 after many of its members joined the armed forces.

While Gordie did not have time to play organized sports, he

spent hours at the nearby YMCA and the State Armory playing basketball. Even as a youngster, he was commended for his on-court skill and his large hands, allowing him to palm a basketball at the age of fourteen.

Once, when the armory was closed, Gordie climbed over the barbed wire fence to let himself and his buddies inside, knowing a secret way to get in. He cut his arm on the wire and walked home a bloody mess. The injury required thirty stitches, and his mom was upset that he bled all over her floor.

Gordon attended Drury High School as a freshman in the fall of 1941. After war was declared, he decided to leave school in his sophomore year and join the Army. Gordie enlisted in April 1942 and was assigned to the 9th Quartermaster Corps at Camp Lee, Virginia. Several months later, he was selected to attend the three-month advanced mechanic's course at Hampton Institute in Virginia.

Upon graduation, Gordon was promoted to Corporal and was stationed at Fort Leonard Wood in Mississippi. During this time period, his brother Quentin and three of his sisters also joined the military in a display of patriotism. Quentin enlisted in the Navy and was stationed in Boston; Marcella entered the Signal Corps and was located in Washington, DC; Mildred qualified for the Nurses' Aid Corps; and Arlene became a private in the Women's Army Corps (WACs). She was stationed at Camp Gruber, Oklahoma.

Gordon was assigned to the 92nd Infantry (Buffalo) Division within a year of his enlistment and was sent overseas to participate in the perilous African campaign. Still working with the quartermaster corps, he was engaged in the dangerous effort to transport munitions and other supplies to the front lines.

The division had previously fought in France during World War I. When the 92nd was reactivated in 1942 as part of the Fifth Army, the unit, out of tradition, retained the "buffalo" patch as its divisional symbol. The American buffalo was selected as the divisional insignia due to the "Buffalo Soldiers" nickname given to African American cavalrymen in the nineteenth century. The division had 15,000 officers and men. Senior officers were white, while some junior officers and all non-commissioned officers and enlisted personnel were black.

In the spring of 1944, the government, responding to pressure, agreed to allow black soldiers to serve as infantrymen, and several months afterward, once the African campaign concluded and replacements were found, the 92nd Division was thrust into the Italian campaign. Assigned to the US Fifth Army, Gordon's unit occupied the westernmost end of the Allied front, while the Eighth Army attacked across the eastern portion of the Italian Peninsula.

Resupply was so difficult in the mountains that roads were impassable to vehicles. At one point, the 92nd employed mules and horses to transport ammunition, food, water, and even anti-tank guns to the beleaguered soldiers. The efforts sustained the Buffalo soldiers, and they continued to advance, liberating towns along their path.

After months of fighting in mountainous terrain against a stubborn enemy, the 92nd was credited with liberating a number of Italian cities, including La Spezia and Genoa. Gordon was in the midst of the assault by the 92nd that advanced hundreds of miles, broke through the Gothic Line, and helped capture thousands of German prisoners. Its soldiers earned over 10,000 decorations and citations, including two Medals of Honor. The cost to the Buffalo soldiers was over 2800 killed, wounded, or captured.

Gordon had spent thirty-six months overseas, earning the European–African–Middle East, Italian and American campaign medals, the World War II Victory medal, the American Defense Service Medal, and a Good Conduct Medal. He also had other uniform badges, including one for marksmanship.

Promoted to Technical Sergeant, Gordon returned home and decided, after three and a half years of Army life, to reenlist and make it a career. He spent a 90-day furlough with his parents, then reported to Fort Devens, Massachusetts, for reassignment to the Pacific Theater.

His new orders directed him to the 24th Infantry Regiment, an all-black unit with the 8th Army. The regiment was initially directed to serve occupation duty in Okinawa while receiving replacements. After this stop-over, the regiment sailed to Japan in early 1947 and then rode a train to their new home at Camp Gifu, a former kamikaze base left in shambles after the war.

Gordie, home with his parents on furlough, before heading to Korea. *[Courtesy of the* Transcript, *August 23, 1950]*

African Americans had served honorably in their nation's military since the American Revolution, but it wasn't until July 26, 1948, when President Truman, after considerable prompting by the black community, signed an Executive Order that the military was directed to end segregation. The order stated, "There shall be equality of treatment and opportunity for all persons in the armed forces without regard to race, color, religion or national origin." (With some reluctance by military leaders, the military was fully integrated by 1954. It took years afterward for uniform acceptance.)

The base facilities at Camp Gifu were eventually repaired, and the base became partially integrated. The Army's supply chain, which had been interrupted when the unit moved, finally caught up with the soldiers, bringing needed clothing and boots. The unit would remain there, continuing to refit and train until their deployment to Korea. Aside from tactical training, another of their objectives was to patrol nearby villages looking for hidden weapons and munitions left over from World War II.

Off-base occupation duty in Japan was easy duty. The local populace was still recovering from the privations of World War II. The soldiers' money was eagerly accepted and could stretch far. Even junior enlisted men could afford to hire servants, and alcohol was cheap. The local black market thrived with items hard to purchase elsewhere.

During this time period, Gordon was temporarily transferred to Yokohama, Japan, for training. As noted in the *Transcript*, Gordon Greene had been "found qualified as a light-truck driver and auto-mechanic, and now entitled to advancement in rank under the Army's career plan, the Army announced. A member of the 577th Ordnance service company, the North Adams man is a member of the occupying forces in that country" (7-31-1948).

Months later, Corporal Greene returned to Camp Gifu and, in early 1950, participated "in winter maneuvers at the base of Mt. Fuji, Japan's most famous mountain," describes the *Transcript* on March 1, 1950. The article reports that he was now an infantry squad leader with the 24th Regiment. With the approaching conflict in Korea and the need for experienced NCOs, Gordon's military job had shifted from the Quartermaster's corps to infantry. Gordon's rank changed when the Army adjusted its enlisted ratings in the early 1950s. Formerly a Technician Sergeant (three stripes) in a support service (the Quartermaster's corps), he became the equal rank of Corporal in the infantry.

While the training was helpful, his unit's equipment was still outdated, in poor condition, and they were provided limited training areas. Experienced officers and NCOs were lacking. It has been reported that some experienced senior white officers were reluctant to serve in all-black units, concerned the assignment would affect their promotional opportunities.

The Korean War started on June 25, 1950, just a few months after the 24th Regiment's Japan winter maneuvers concluded, when tens of thousands of North Koreans equipped with tanks and artillery invaded South Korea.

Gordon's unit, the 24th Infantry Regiment, also known as the "deuce-four" part of the 25th Infantry Division of the Eighth Army, was quickly dispatched to Korea in late June 1950. His parents said Gordon's last letter was dated July 7, 1950, telling them that he had arrived in Korea after a four-day train ride from Camp Gifu to a port on Japan's west coast. He had been transported by boat across the Sea of Japan to Pusan, Korea.

[Courtesy of the author]

The 24th was not prepared for combat, but it was one of the units closest to Korea when war was declared. It was under-strength, composed of many recent draftees with not enough training. A sprinkling of stable, solid veterans like Gordon comprised its backbone, but not enough to make a difference against a battle-hardened group of the North Korean People's Army (NKPA).

They were given no chance to ease into combat; the undertrained, ill-equipped regiment was thrown directly into the firing line alongside the retreating South Korean forces. With no tanks, and few anti-tank weapons, the 24th was pushed south by well-trained communist forces. The shock of direct combat, heat, humidity, mountainous terrain, and the stench of warfare was overwhelming to the young foot soldiers. The first few months of conflict saw tremendous casualties for both the Americans and the South Koreans. By the end of the Korean War, the United States would lose 33,700 men to combat fatalities.

Corp. Gordon A. Greene Dies of Wounds in Korea

Veteran of North Africa, Anzio Beach and Italian Campaigns Second North Adams Boy to Give Life in Far East Conflict — Son of Mr. and Mrs. Anthony M. Greene Nine Years in Army.

A North Adams boy who fought in North Africa, at Anzio beach, and throughout the rugged Italian campaign in World War 2 has lost his life in Korea.

Died July 31

Corp. Gordon A. Greene, 26, died on July 31 of wounds recieved in action with the heroic 24th Infantry regiment, according to an Army department telegram received last night by his parents, Mr. and Mrs. Anthony M. Greene of 26 Chestnut street. He was this city's second Korean war casualty reported in as many days, the death of Corp. Patrick J. Gaffey of 144 Crest street on Aug. 8 having been announced yesterday.

The Army department telegram gave no further details of Corp. Greene's death, but his parents said their last letter from him had been received on July 7, telling of his arrival in Korea from Japan. Since he was a member of the 24th Infantry regiment, they believe he probably had been with the Negro battalion outfit in fierce combat around Chinju on the far southern flank of the U. S. defense

Gordie's hometown paper breaks the tragic news. *[Courtesy of the* Transcript*]*

A little over three weeks after arriving in Korea, Gordon was killed.

It was July 31, 1950, the date aligning with the timeframe when the entire Southern Korean front was ablaze with communist offensives and renewed American counterattacks. North Korea's KPA (Korean People's Army) infantry divisions decimated several US regiments, including Gordie's, near the town of Chinju.

The units often fought for survival against a numerically superior force, incurring thousands of casualties. The retreat was so chaotic and the information so sketchy that Gordon's parents did not receive an Army telegram until August 22, 1950, notifying them of his death weeks earlier. As his grieving, devastated father read and reread the telegram, he was remembered as saying, "I guess his luck just ran out."

In a matter of several months, Gordon's unit became greatly reduced. Later articles from the local newspaper state that "[Gordon] was fatally wounded while fighting with the heroic 24th Infantry regiment and was a member of the Negro battalion that suffered terrific losses in the early fighting around Chinju where the United Nations forces were in danger of being driven back into the sea" (*Transcript* 08-27-1951).

Gordon was twenty-six years old when he was killed, just after he had decided to make the Army his career. He had already served nine years, surviving months of combat in World War II from the North African campaign to the Anzio beach landing and onward to the Italian offensive. It was a tragic ending to his service.

In mid-August, Gordon's body arrived in San Francisco aboard the transport *Lynn Victory* with the remains of 505 other war dead.

On August 28, 1951, Gordie arrived in North Adams aboard the *Minute Man* train, which ran between Boston and Troy, New York. The soldier was "met by his grief-stricken family and representatives of local veterans' organizations and service units. Present when the flag-covered casket was carried off the train were Cpl. Greene's mother and father, Mr. and Mrs. Anthony M. Greene, his sister, and a brother-in-law." In addition, USO, Army/Air Force recruiter, and VFW representatives were there to greet the coffin (*Transcript* 8-28-1951).

Corporal Greene's body was accompanied by an Army sergeant from the same 24th Infantry regiment, then moved from the train to a hearse sent by Louis San Soucie funeral home.

The funeral was held the next day, August 29, 1951, at St. John's Episcopal Church, where the young man had attended services with his family. The body was escorted to the family's plot in Eastlawn Cemetery in Williamstown. "A group of six negro members of a military police unit from Westover field (firing squad) and eight members of Company K, Massachusetts National Guard will participate in the burial services (honorary bearers)," wrote the *Transcript*.

On June 25, 2000, the 50th anniversary of the beginning of what has been called "The Forgotten War," North Adams City Hall sponsored a dedication ceremony to the twelve local men who died in the Korean War. The ceremony began with raising the flag, and a former chaplain opened with a prayer. The mayor reflected on their honored service. The *Transcript* reported that an elderly Gold Star mother tearily stated, "I still miss my son . . . ."

Alex Daugherty, Gordon's nephew, attended the anniversary

ceremony on behalf of his family, "graciously accepting the memorialization of his uncle," reports the *Transcript*, and saying, "My only regret is that I didn't have long enough (time) with him as an uncle to grow up with . . . but he is still missed." Alex now has his uncle's purple heart medal, and as is the custom, Gordon A. Greene's name was inscribed on the back of the vaunted award under the inscribed words "for Military Merit."

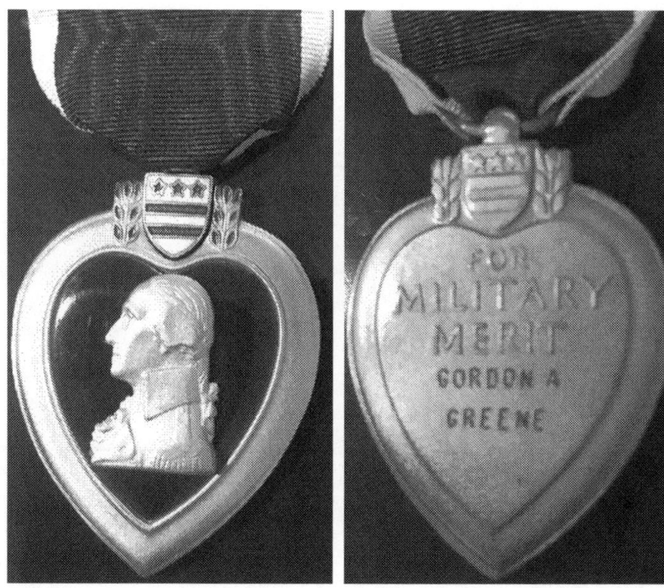

[Courtesy of the Greene family]

The picture that accompanies Gordon's obituary in the *Transcript* shows a serious, solid, strong-looking man with a hint of a mustache. The picture was taken shortly before he left for Japan in 1947. He is dressed out in full uniform regalia, with corporal chevrons, two rows of campaign ribbons, sleeve hash marks indicating years of service, and eight overseas bars, each representing six months. Gordon's decorations would finally include a posthumous purple heart, the Korean Service and Campaign ribbons, and the Republic of Korea Presidential Unit Citation.

Not long after his death, the City of North Adams honored

Gordie by naming Greene Street after him. One of his sisters named her son Gordon, a living testament to his vibrant memory.

This twenty-six-year-old career soldier faithfully and successfully served in the US Armed Forces during a difficult and racially charged time for black Americans. Gordon Anthony Greene died side-by-side with his comrades helping to secure the freedom of another nation. Both nations are indebted to him and to his family for this, the greatest of sacrifices.

Sergeant Peter W. Foote

Silver Star

VIETNAM

As he led his unit up Cemetery Hill, the machine gun, mortar, and rocket fire was deafening. Shrapnel and bullets filled the air like angry bees. He remained in front, choosing unobstructed pathways up the hill. It was early morning on January 30, 1968, the opening moments of Vietnam's Tet Offensive. The enemy had taken control of the hill, with the goal of overrunning and capturing nearby Tuy Hoa City, the province capital. It was Peter Foote's job to get it back.

It was not many months before that Pete had been quarterbacking the University of Massachusetts freshman football team, and not long before that he'd been hard at work leading the Drury High School Blue Devils. Again, to deafening noise . . . albeit much friendlier.

PETE GREW UP IN North Adams, Massachusetts, on 22 Lawrence Avenue, the son of Edward and Claire (Roberts) Foote. For a number of years, the family had used the last name *Frappier*, of French origin, but Pete was always Foote.

His parents were both graduates of local Drury High School, and his dad Edward served as a radio operator in the Marines, stationed on several air bases during World War II. The young couple married not long after his return; Peter was born in 1947 and his only sister Paula in 1954. Ed worked many years in the drafting department for the General Electric Company, and Claire was a homemaker. The family were communicants of Saint Francis Church.

The Footes lived not far from Mark Hopkins Elementary School, which Peter attended from kindergarten through eighth grade. He belonged to Mark Hopkins' Cub Scout Pack 99 and, in an early sign of his athletic ability, won the sack race for eight-year-olds. Pete and his best friend and nearby neighbor Joe Ames enjoyed many a day of pick-up sports at the school playground.

Pete's dad, formerly a player on the St. Anthony Crusaders semi-pro football team, appreciated and promoted Pete's budding ability, although at times, he couldn't resist teasingly describing Pete's prominent ears as looking "like a taxicab with the back doors open." Pete had an infectious personality and shook off the teasing easily. He wanted to have fun, was sometimes a practical joker, and loved a good laugh. People enjoyed being with him.

Often Joe and Peter would go to Fischlein's, the nearby grocery store, and charge two half-gallons of Neapolitan ice cream to Pete's parents' account. They would return to Pete's house, grab two spoons, and relax on a nearby hill at North Adams State College while devouring their treasure.

Pete's love for ice cream was legendary. One summer when the Footes vacationed at Hampton Beach, a local restaurant was offering a challenge to anyone who could eat their large banana split known as the "White Whale." If a person could eat it all, he didn't have to pay for it. Pete accepted the challenge and promptly ate all 12 scoops of ice cream with assorted toppings in record time. The feat was cheered by other patrons, with an ovation to the kid with the big smile, crew cut, and prominent ears.

Pete and Joe liked to fish when they were not playing sports. They would rent a rowboat and fish at Cheshire Lake or tromp the Deerfield River and Tunnel Brook, looking to catch elusive trout. But Pete's athletic ability soon became evident when he played Little League baseball on the police department team. He was designated as the starting left-handed pitcher and had a dominant curve and fast ball. He was also a powerful left-handed batter.

In addition to baseball, Pete played YMCA basketball in the

local Saint Francis Church League, where his long arms and large hands were advantageous in rebounding and defense. Pete relentlessly played football, both intergrade and sandlot, always trying to perfect his spiral passes. Usually with Joe as the receiver, Pete would pass between telephone poles, with the poles marking first downs.

He was most popular with the neighbor kids, offering them rides on his light-colored Vespa scooter, his unpretentious around-town transportation. The kids would gather at his house, and his dad would help line up the waiting riders. One of his frequent passengers was a young next-door neighbor, an altar boy at Pete's church, St. Francis.

It seemed like he and his friends Ray, Joe, Butchy, Billy, Tommy, and Kevin were always playing or practicing sports either at school or nearby playgrounds. When not practicing, they could be seen with their friends hanging out on the corners of Eagle Street or at Apothecary Hall on Main Street, usually discussing sports, girls, and cars.

Eventually growing into his 200-pound, 6 foot 3 inch frame, Pete was well suited to play high school sports, and his abilities blossomed when he attended Drury High School in the early 1960s. Pete, as a three-sport athlete, was one of the top high school basketball players in western Massachusetts. As co-captain, forward, and center, he led his team in scoring and rebounding during the 1963–64 season. He and his teammates thrilled the community with their exciting games at the North Adams Armory.

Drury had a strong basketball team. They'd had several 17–3 seasons. One of their biggest wins was at the Western Mass Schoolboy Championship Tournament. With no divisional seeding, the Blue Devils beat the much larger school, Springfield Cathedral, in overtime, finishing at 91 to 88. Pete scored twenty-five points while bothered by a pulled ligament in his knee that later required surgery.

Pete also pitched and played first base for Drury Blue Devils baseball, although his renown came through football. When merely a sophomore, he quarterbacked the Blue Devils football team.

The sports accolades continued to pile up. During his four high school years, Pete was chosen by the magazine *High School Basketball Illustrated* as one of the region's top players. He was also awarded the Rev. Theodore Hesburgh plaque as the county's outstanding scholar-athlete for 1965. He received All Western Mass recognition for basketball and football and was named athlete of the year by a local radio station. Berkshire County Basketball Officials Association awarded him their Sportsmanship Trophy.

PETER WELLESLEY FOOTE
22 Lawrence Avenue
Pete . . . Bob Cousy . . . Mirror, mirror, on the wall, of course I'm the fairest of them all!!! . . . loves to wear Bermudas.
V. Basketball 1,2,3; (Cap't) 4; V. Baseball 1,2,3,4; Varsity Club 1,2 (Sec't.) 3,4; V. Football 2,3,4; Nu Sigma (V. Pres.); Prom Comm. 3.

Peter Foote in 1965 Drury High School yearbook. *[Drury High School]*

At the time, the local newspaper wrote of Pete: "that talented player throws, runs and calls signals to meet any coach's standards."

His senior yearbook lists him as playing varsity basketball and baseball for four years and football for three years, and notes his membership and vice-presidency of the Nu Sigma Epsilon society, an honor group at Drury. The yearbook also says he "loves to wear Bermudas."

As high school graduation approached, Pete was courted by a number of colleges to play football and was selected as an alternate for the West Point Academy. He decided to attend the University of Massachusetts on a full football scholarship and competed with other student athletes for the school's quarterback position.

In Pete's freshman year, Ed Foote traveled with Pete's best friend Joe to watch Pete play in the annual spring inter-squad scrimmage. The varsity's current quarterback, Greg Landry, an eventual professional football player, called the plays the first scrimmage and his talent was very evident. Pete ran the next scrimmage and efficiently threw some great passes, working his team down the field. The crowd was impressed, and Joe could see they considered the talented eighteen-year-old a very possible replacement after Landry graduated.

At this point, something occurred in Pete's life that caused this promising young man to quit college and join the Army after his freshman year. His definitive reason is lost to history. Some assume it might have involved the loss of a girlfriend; others chalk it up to coming from a patriotic family (his father had been a Marine in World War II). Pete would often discuss the military with his barber, a friend who had served as a paratrooper in World War II (in fact, in the same regiment where Pete ended up).

It's notable, in any case, that this accomplished young man abandoned the sport he loved and the state of his birth to enlist in the US Army in September 1966 with the one desire to become a paratrooper. Surely he knew that after training he would be deployed to Vietnam.

Peter completed his physical and headed to basic training in September. After the completion of basic, he attended Advanced Infantry Training at Fort Gordon, Georgia, and then moved onto Airborne Training at Fort Benning, receiving his paratrooper wings that next February. He went home on furlough; then, after a little more than six months in the service, he found himself in Vietnam on March 29, 1967.

After completing a brief period of additional jungle and radio operator training, he was assigned to Company D, 4th Battalion, 503rd Infantry Regiment, 173rd Airborne Brigade. His Military Occupational Specialty was 11B2P-Infantryman-Parachute Qualified.

His unit, the 4th Rifle Battalion, was different from other infantry units. His battalion remained "in the bush" for weeks at a time and never had the luxury of returning to a forward base for rest and refitting. For the 4th, there was no stand-down; they stayed in the jungle their entire time. Every fifth day, helicopters would ferry C rations, ammunition, grenades, and a lukewarm meal to them so they could remain in the action.

With the unit spending all its time on-line, casualties were considerable, and numerous replacements for those killed and wounded would be offloaded from the resupply helicopters.

The unit aggressively sought out the enemy in search-and-destroy missions or as a reaction force, responding to unanticipated assaults. They worked and slept in the open under arduous conditions. The unit patrolled and fought in triple-canopy jungle or feces-saturated rice paddies, often sleeping in foxholes half-filled with water. They lost weight and many developed a pallor from working in the jungle with little or no sunlight.

They wore load-bearing harnesses across their shoulders, carrying items needed in case of immediate action, including rifle magazines, a number of grenades, canteens, a bayonet, and a first-aid kit. They carried everything they owned on their back, with rucksacks weighing in at 80 pounds or more. The rucksacks held more rifle magazines, fragmentation and smoke grenades, probably a Claymore mine, a poncho, a camouflage nylon liner, C rations for 5 days, and minimal hygiene items. The troopers seldom carried any extra clothing; most did not wear socks or underwear due to the high humidity, and their torn and rotten fatigues were replaced sporadically on resupply runs.

Pete had found himself assigned to a premium fighting unit, albeit one with the highest casualty rates in Vietnam. Over a one-year period, out of the 750 soldiers in the battalion, 125 were killed and over 500 wounded (some repeatedly). The 4th Battalion's biggest battles were at Dakto in the Central Highlands near the Laotian/Cambodian border and the 1968 Tet Offensive fought near the city of Tuy Hoa (Tooy Wah) in the coastal province of Phu Yen.

The unit fought in Dakto several different times when large incursions of enemy soldiers were coming across the border. Fighting occurred in the summer and fall of 1967 in Dakto, then the 4th Battalion was moved to Tuy Hoa to protect the area's rice harvest from being seized by the enemy.

The unit returned to Dakto in the fall of 1968, when four regiments of North Vietnamese were infiltrating into South Vietnam to push American forces out of the Central Highlands. The 6000 insurgents were met by Pete's 173rd Airborne Brigade. A series of vicious hill battles were fought in November. It was in the worst terrain possible, with unending steep ridges and triple canopy jungle in one of Vietnam's most remote areas.

Company D fought and won the battle for Hill 823 on November 6, and two weeks later was fighting for survival in a five-day battle for Hill 875. The US losses for the hilltop battles were 361 killed and over 1400 wounded. North Vietnam losses were estimated at over 1600. The 4th Battalion had 28 killed and 123 wounded. One of the wounded was Pete.

On November 20, just as his unit reached the top of Hill 875, Pete suffered neck shrapnel wounds. He was treated quickly by a medic and sent to a hospital for brief recovery. Hill 875 was eventually secured by American troops on Thanksgiving Day. Pete's parents and Joe had watched the Dakto battles on TV, hoping to see him and relieve some of their anxiety. His parents were subsequently notified of his being wounded. For its actions, the 173rd Airborne Brigade was awarded a Presidential Unit Citation.

In December, Company D, including a recovered Pete, returned the three hundred miles to Tuy Hoa and continued participating in the ongoing Operation Bolling, resuming search-and-destroy missions and providing security for rice gathering. Tuy Hoa, which translates in English to "endowed with tranquility," was on the coast. The troops were happy to be away from the difficult Central Highlands where mornings and nights were cold.

Pete was a highly regarded squad leader. He had a convincing rapport, was always willing to help, and people listened to him. One squad member said, "Pete took care and watched out for his men, especially the replacements." He wished Pete could have been his squad leader the whole time he was in Vietnam. He also recalls Pete feeling sorry for a small Vietnamese boy struggling to walk in the mud. Pete picked up the boy and placed him on top of his 100-pound rucksack, carrying him on his back most of the day.

"This is the kind of guy Pete was," recalls the squad leader.

When the Tet Offensive commenced, early morning on January 30, 1968, Pete's Company, designated a reaction force, was confronted by the NVA's 5th Battalion of the 95th Infantry Regiment. The enemy troops had surged across the province with the intention of capturing Tuy Hoa City and had some limited success holding on to a few areas.

Company D, decimated after Dakto, now numbered 83 troopers. (By this point in battle, Companies C and D assisted each other as needed; some reports list Pete serving in C and some in D.) The company was helicoptered to battle and began a series of skirmishes near a local prison and an artillery battery to push out insurgents. The action moved to an area called Cemetery Hill. Since hostages were involved, it would be a fight without an advance artillery barrage or air support . . . just rifles, bayonets, and grenades.

A point element led by Pete's reduced squad started climbing through a wide-open space near the cemetery. They presumed the enemy was dug in at the top of the knoll and had an elevated field of fire for machine guns and mortars. Tree snipers and soldiers were concealed in spider holes all along the way. The fighting was fierce and, at times, point-blank.

Pete, carrying his M-16 and an M-79 grenade launcher, quickly eliminated several sniper positions while under fire. As he continued uphill, he spotted a seriously wounded radio operator needing immediate help. Rushing to him, Pete was able to recover the trooper and drag him back out of the line of fire.

While rendering critical first aid, Pete's six-foot-three height, even while kneeling over the grievously wounded trooper, was just tall enough of a silhouette that a sniper fired and hit him in the head; he gasped, fell over, and presumably died instantly. One of Pete's squad members who was close by was just about to warn him to get down when the bullet hit.

Several months previously, Pete had saved this person's life when a sniper laid a trap with what appeared to be an abandoned antique French rifle. As the trooper bent over to pick up the souvenir, Pete shoved him to the ground, and the sniper's bullet hit the dirt near his elbow, barely missing him.

The sniper who killed Pete and shot at his squad member was probably using a Russian, five-shot, bolt-action Mosin-Nagant rifle. The rifle was cheap to purchase, rugged, and reliable in all conditions. In the hands of a good sniper, the scoped 7.62 X 54R cartridge speeding at 2600 feet per second was accurate up to 800 yards, a risk Pete would have known even as he tended to his countryman.

After Pete's death, the remaining troopers surged through the cemetery, retook the ridge, and pushed out the remaining insurgents. Later the troopers returned to retrieve their dead and wounded to be placed on medevac choppers.

Nineteen of the eighty-three members who assaulted the hill died, and almost everyone else was wounded. The troopers of Companies C and D became some of the first fatalities of the Tet Offensive. From an initial company roster of 150 soldiers, fewer than a dozen were able to participate in a memorial service held several days after the action. The entire 4th Battalion was awarded the "valorous unit citation" for this action.

It is estimated that over 200 insurgents were killed and buried in a mass grave. The NVA's 5th Battalion became ineffective at a high cost.

For Pete and his family, it was the highest cost. When Pete died, he earned his third Purple Heart. The 4th Battalion would often transfer those who had been wounded three or more times to a safer job in the rear for the rest of their tour of duty. Had he survived the final injury, it's likely he'd have made it home.

Not long after the battle, Pete's distraught parents were notified by Army representatives and Western Union telegram of their only son's death. As a friend later said, "When he died, he brought the war home to North Adams." The city was profoundly chastened by the loss of

A framed collection of Pete's medals and unit insignias is displayed in the North Adams skating rink dedicated in his honor. *[Mike Remillard Photos]*

its well-known and well-liked star athlete. Pete had been scheduled to be home in six to eight weeks. His buddy Joe remembers receiving a letter from Pete saying he would be home in fifty-eight days.

When his body was received at the funeral home, Pete's parents talked about verifying that the body was really him. So, Joe and Pete's grandfather met with the funeral director to discuss opening the sealed, aluminum casket. The casket had been tightly sealed for almost two weeks, and the director discouraged them from opening it. Although dental records verified it was Pete, they never got to look for the scar on Pete's knee from a recent operation.

The funeral was held downtown at St. Francis Church on February 16, 1968. Soldiers from Fort Devens provided an escort and honor guard. The North Adams mayor was present, and Pete's coaches, classmates, and family friends filled the church, with his teammates as honorary pallbearers. His young neighbor, the former Vespa passenger, served as one of Pete's altar boys. Burial was held on a cold day at Southview Cemetery with full military honors, including "Taps," rifle volleys, and a flag presentation to Mrs. Foote.

Shortly after the funeral, Joe co-founded the Peter W. Foote Memorial Scholarship program. The memorial awards college scholarships to the most deserving senior boy and girl athletes at Drury High School. It has been helping athletes since 1968.

On May 15, 2010, over forty years after his death, the City of North Adams also recognized his heroics and sacrifice by renaming the local ice-skating rink "The Peter W. Foote Vietnam Veterans Memorial Skating Rink." Over a hundred people, including a color guard, came to the dedication. The Master of Ceremonies, his friend Joe, pointed out that Pete now rests in Southview Cemetery about 300 yards north of the rink, close by his legacy.

Adults, children, and local hockey leagues all enjoy visiting the skating rink named after Pete. *[Mike Remillard Photos]*

Although Pete's parents and sister have now passed away, and college buildings occupy the site of his former home, Pete lives on in the memories of many sports fans and friends, on the building that memorializes his name, and through his scholarship program.

As one of his friends stated, wistfully, "You just knew he was going to be successful, and the world was a better place with Pete in it."

Postscript: *Pete Foote's personal military decorations include the Silver Star, a Bronze Star, three purple hearts, the Vietnamese Cross of Gallantry with Palm, plus a number of unit and tour awards.*

> *AWARD of the SILVER STAR (Posthumous)*
>
> *"...Company C was engaged in heavy combat with an estimated North Vietnamese Army battalion and was receiving automatic weapons and sniper fire. Specialist Foote was serving as point man for the lead element and immediately engaged the enemy with effective fire from his M-79 grenade launcher. He continued to fire, killing several of the enemy, until he ran out of ammunition. At this time, the lead element was ordered to move left. Specialist Foote again took the point and led his platoon across the enemy's front. When he reached the left flank, Specialist Foote saw a wounded man crawling back toward friendly positions and, with complete disregard for his own safety, ran forward to aid him. As he ran to the casualty, he was wounded severely in the leg, but in spite of this continued forward. Upon reaching the wounded man, Specialist Foote discovered that he was hardly breathing and although completely exposed to enemy fire, he administered artificial respiration to him and began dragging him back to the safety of friendly lines. He successfully reached the friendly position, but while placing the wounded man behind cover, was fatally wounded by an enemy sniper. By placing the safety of his fellow man above his own, he had saved the life of another. Specialist Foote's outstanding display of personal courage and devotion to duty was in keeping with the highest traditions of the military service and reflects great credit upon himself, his unit, and the United States Army."*

Specialist Michael R. DeMarsico II

Bronze Star

AFGHANISTAN

Panjwai District in Kandahar Province is one of the most volatile areas in Afghanistan. The villages in the area are collectively known as the "Horn of Panjwai," considered the birthplace of the Taliban, a longtime insurgent stronghold. Over the years of Operation Enduring Freedom, many battles have been fought there. It is an area infested with improvised explosive devices (IEDs). During the war in Afghanistan, most Canadian casualties occurred in Panjwai.

One of the Taliban-controlled villages in the region is called Zangabad. That is where Company C, Michael DeMarsico's company, found itself in August 2012, in a heavily fortified structure called Forward Operating Base (FOB) Zangabad.

MICHAEL'S MOTHER RECALLS WHEN he was nine, in September 2001, watching television coverage of the attack on America and saying to his mom that he wanted to join the military to protect the United States. As moms do, she reaffirmed him, and several minutes later, he returned and said, "but if there is a war, I would like to come home."

Little did they know that just shy of ten years later, Michael DeMarsico would be in the Army, involved in America's longest war, serving in Afghanistan as an infantryman. During the waning years of Operation Enduring Freedom, Michael's brigade would continue fighting to contain the Taliban and crush its support for al-Qaeda.

For Michael and his squad of infantrymen, every daily foot patrol in the Panjwai district had a chance to become deadly. Local tribal sympathies rested with the Taliban, undetectable IEDs were everywhere, and soldiers were targets of opportunity for any nearby snipers.

But Michael's story starts many years earlier, in North Adams. He was born at the North Adams Hospital on December 22, 1991, to Michael R. and Lisa (Babcock) DeMarsico.

Lisa loved working with children and, for many years, worked in a special behavior program; she continues today to work as a preschool teacher. Her husband Michael works as a maintainer at the nearby college and is an accomplished woodworker. The family would have five children, three girls and two boys, the older of whom was named for his father.

Michael's maternal grandparents were of French ethnicity; the grandfather worked in construction while the grandmother was a homemaker. His paternal grandparents, both of Italian descent, were longtime employees of the Sprague Company. Michael's paternal great-grandparents had emigrated from Italy via Ellis Island.

The DeMarsico family lived on Bracewell Avenue in a two-apartment duplex. Michael attended Sullivan Elementary from kindergarten through the sixth grade and Conte Middle School through the eighth grade. As a little tyke, he always played with green plastic toy soldiers and trucks and talked about wanting to be a soldier. When a bit older, he began playing with World War I and II toy airplanes and reading history books about the same era. He was fascinated when watching the History Channel.

Michael also loved being a member of St. John's Cub Scout Pack 35, where his dad was cub leader. The highlight of his membership was using a kit to create a sailboat and racing in the annual "Rain-Gutter Regatta." The self-constructed sailboats would race down rain gutters (like those found on a house), powered by each cub's breath. Michael always had the faster boat until Adam, his brother, finally beat him one year.

The boys also had an annual pinewood derby competition, in which a block of wood, four wheels, and four axles would be hand carved and designed. The wooden racers competed against each other. Michael loved these competitions.

His dad coached Michael in T-ball and Little League, and Michael went on to play in the older Babe Ruth League. He played the center field position, but on occasion he was pressed into action to pitch for his team. He also played for the LaFesta Baseball Exchange, a local group of baseball players that scheduled rotational summer games with a team from Boston's North End.

As a youth, Michael played Pee Wee football. In his freshman and sophomore years at Drury High School, he was a running back for the Blue Devils. He completed his junior and senior years through an off-campus program at Berkshire Community College, achieving honor roll each year. While attending the off-campus program and interning with security, he found a passion for criminal law, an area in which he hoped to work one day.

He loved nature and was adventurous, and most weekends he could be found with friends hiking trails in the surrounding mountains, climbing Mount Greylock, or visiting Natural Bridge. Happiness was making campfires, hanging out with friends, and enjoying the outdoors.

He graduated in June 2010 and spent the next several months considering which branch of the military to join. During this time, he worked part-time at Boston Seafood Restaurant and bicycled at night to Pittsfield to visit his girlfriend. His parents thought it was starting to look like a serious relationship; the trip to Pittsfield and back was fifty miles.

Michael's enlistment in the Armed Forces began in February 2011. He decided to join the Army in the profession he always had desired, as an infantryman.

He was designated an 11Bravo and headed to Fort Benning, Georgia, for what the Army calls its One Station Unit Training (OSUT), which combines ten weeks of Basic Combat Training (BCT) with approximately 12 weeks of Advanced Individual Training (AIT).

Sending home a few words of love to the family. *[Courtesy of DeMarsico family]*

Soldiers remain together throughout the entire time, with the same drill instructors streamlining the learning cycle and enhancing unit cohesiveness.

In BCT, Michael learned Army fundamentals such as core values and heritage while also being trained in practical things relating to his job as an infantryman, such as marksmanship, map reading, hand-to-hand combat, and tactics.

Michael sharpened those skills in AIT and became an expert in others, such as reconnaissance, advanced tactics, and fieldwork.

During his time at Fort Benning, Michael received the nickname *DeMar.* When he and another soldier had similar last names, the drill instructor said he needed a way to tell them apart, so he started using *DeMar.* From then on, his military friends would only know him by that name.

Michael graduated from basic training and AIT at Fort Benning in June 2011 and initially had orders for a unit in Hawaii but requested instead airborne training. He was waitlisted for an airborne position and then selected to join a mechanized infantry group at Joint Base Lewis–McChord (JBLM) in Washington State.

Joint Base Lewis–McChord was a merger in 2010 of the Army's Fort Lewis and the Air Force's McChord Air Base. It is a training and mobilization center for all services, intended to train, support, and deploy US combat units. Its geographic proximity means that units can deploy by air and equipment by water.

As an 11Bravo at Fort Lewis, Michael joined the 1st Battalion of the 3rd Stryker Brigade Combat Team, a mobilized infantry force composed of over 3000 soldiers and named for its use of

Stryker vehicles. These combat teams of motorized infantry were well-trained and capable of sustained independent operations against insurgents.

The Stryker Brigade received its name from the versatile, 19-ton Stryker Infantry Carrier. Armed with a .50-caliber machine gun and/or a Mark 19 grenade launcher, it has four hatches on top for the driver, gunner, and two more personnel, allowing for 360-degree coverage.

This durable "taxi of war" could ferry soldiers while protecting them from most IEDs, shrapnel, and bullets. Besides serving as an infantry carrier, it would be used to medevac the wounded and provide support fire for troops in action or reconnaissance. An eight-wheeled Stryker has two crew members and carries a squad of nine. This large, ponderous vehicle can even attain speeds of 60 miles an hour on hard-packed roads.

DeMar's platoon had four Strykers. Company C (Comanche) had approximately twenty of these behemoths. Michael settled in with Comanche and quickly made close friends.

This photo taken by Michael near Mount Rainier in Washington holds a place of honor in in the family home. *[Courtesy of DeMarsico family]*

He and his buddies hiked in the nearby mountains and enjoyed camping in the lava beds at Mount St. Helens. Photographs show them trying a bit of snowboarding. Michael sent home several

DeMar in the early stages of training. *[Courtesy of DeMarsico family]*

photographs of the mountain ranges and lakes. Even today, several framed pictures hang in the hallway of the family's home in North Adams.

Training had been ongoing since Michael arrived at Lewis–McChord in July 2011. Company C conducted individual skills training with M4 assault rifles and machine guns and then worked on navigation skills, team/squad/platoon drills, and live-fire exercises.

More extensive exercises began when the 1st Battalion moved to the National Training Center at Fort Irwin, California, for two weeks of unit operations and live-fire exercises.

They returned to the Yakima Training Center in Washington for more training and maneuvers using their Stryker vehicles and live-firing their weapons, the .50-caliber machine guns, and the Mark 19 grenade launchers.

Deemed combat-ready, the unit deployed in March 2011 from Fort Lewis by commercial aircraft. After several connections, the unit entered Afghanistan via Kurdistan on a C-130 and landed at Kandahar Airfield (KAR).

For the next week at KAR, the unit prepared for conditions in Panjwai, where they were to be assigned. They attended classes on IED awareness, first aid, and local culture and spent time on shooting ranges, zeroing in their weapons. The battalion's companies were moved by helicopter to their respective operating bases, which would serve as their home-away-from-home for the next ten months.

Panjwai in southeast Afghanistan has relatively mild winter months, but from late May through August, it becomes scorchingly hot and unusually sunny. It's a desert-like environment. Residents live in mud huts, baked in the sun, usually with roofs that are a combination of mud, sticks, or tarps. Lush green grapes and vines are everywhere during the growing season, decorating mud walls.

The battalion's Area of Responsibility included villages in the "Horn of Panjwai," areas long considered Taliban strongholds. Michael's Company C of approximately 150 soldiers was to be located at FOB Zangabad. Within each company are four infantry platoons. Michael was assigned to the 2nd, with a platoon nickname of "Deuces Wild."

DeMar flashes the platoon sign for Deuces Wild. *[Courtesty of DeMarsico family]*

FOB Zangabad was large enough to shelter the entire company, protecting them with Hesco barriers, extra-large, heavily woven, back-hoe, gravel-filled containers. The thick barriers acted like sandbags, stacked high enough to protect Michael and his company and topped with concertina wire. Around the clock, heavily armed soldiers manned guard towers and entry control points.

Within the confines of the camp, soldiers lived in squad-size tents erected on wooden platforms with air conditioning. Sleeping arrangements were usually bunked twin beds with personal items in duffle bags on the floor underneath. Weapons, ammunition, and grenades were always close by, within arm's reach.

Nearby, truck semi-trailers with metal steps provided shower stalls and sinks for wash-up areas. Porta-Potty clusters were easily accessible. The nearby Morale Welfare Recreation (MWR) tent offered around-the-clock access to phones and computers with internet access. Another small trailer allowed purchases of sundries and hygiene items.

With shortened deployments of ten months, there was no R&R

break. The comings and goings would be too disruptive to unit cohesiveness. Soldiers' free time was limited to watching movies on computers, playing cards, or working out. Michael enjoyed the movies, the gym, and hanging with the guys. His "Planet Fitness in the desert" consisted of some free weights outside in the heat under camouflage netting, on stained plywood.

Squad-size patrols (twelve soldiers) out of FOB Zangabad occurred almost daily. Over half of them were on foot, or what is called *dismounted* patrols. The remainder were ferried to their destination in Stryker vehicles. Squad patrols remained within a four-to-five-mile radius of their base.

The soldiers were heavily armed with M-4 assault rifles, grenades, a grenade launcher, a machine gun, and flash bangs. The squad leader packed C4 explosives. Each person wore a helmet, body armor, and a groin protector. One squad member always carried a mine detector.

Counterintuitively, these heavily armed patrols, which often met and shared tea with village elders, were intended to win the "hearts and minds" of local civilians. Soldiers would offer help, while at the same time seeking intelligence on Taliban movements.

The sweltering soldiers were never quite sure where the villagers' sympathies lay. Could they be friendly because they want something? Or because they are gathering intelligence for the Taliban? Might they ignore requests from perceived military invaders, or is it the Taliban who are perceived as invaders?

During their daily patrols, each squad's every movement was led by one of its members using a mine sweeper in hopes of clearing the way of IEDs, one of the signature weapons used by terrorists. In Afghanistan, from 2001 to 2021, IEDs represented 66% of US casualties. The Taliban also targeted public spaces, houses of worship, and public gatherings, causing 21,000 civilian casualties.

Signature Taliban IEDs used a plastic container the size of a quart of oil, packed with 1.5 pounds of an explosive mixture of ammonium nitrate and powdered aluminum with a detonation cord. The container would be triggered by pressure-plated wooden boards taped together with foam and a carbon rod that activated the device using the least metal possible to avoid mine detectors. The entire apparatus was usually energized by a hidden wire connected to batteries, which were usually replaced late at night or in the early morning hours. Often, cords and batteries would be cleverly buried and hidden around nearby walls and thresholds.

At times, mine sweepers would not pick up the IED signature but human intuition would. A soldier's experience was key. He might notice that something seemed incorrect or out of place, or surmise this would be a good spot for an IED. He might just see a simple ground disturbance.

IEDs were so dangerous they caused patrols to walk slowly in a single file versus spread out in typical patrol fashion. Squad leaders will tell their men to always walk in the footprints of the man in front of them—never deviate. The minesweeper always walks up front, and patrols don't go anywhere without them. Once found, IEDs are flagged, and the EOD (Explosive Ordnance Disposal) team will usually detonate them.

One of the squad leaders within Michael's platoon was adamant about IED protocol since he had been in an explosion that flipped his vehicle. In his case, it involved a larger 200-pound explosive, which was command-detonated (activated by an insurgent located nearby).

During one of Michael's missions, he and his squad discovered a large hut that appeared to be an IED production center. They uncovered explosive powders, carbon rods, wooden boards, and several AK-47 assault rifles. Later, the team would be honored for this discovery, and Michael received the Army Commendation Medal.

It was close to this time that Michael let his mom know that his tour was almost half over. He would see her at Christmas. He was still thinking about what he might do after he got out of the military. He asked his mom to check the availability of criminal law courses at Massachusetts College of Liberal Arts (MCLA) in North Adams.

During Mike's night security watches, he often thought about home and shared with his mom that once while standing watch while his unit slept, he saw flares in the distance, and for a brief moment, it was almost like being home with family watching fireworks on the Fourth of July.

The daily grind of patrols continued, and on August 16, 2012, Michael's squad was headed out on a mission outside the village of Girandai. The mission's primary purpose was to cut down some trees that had been obscuring the view from a surveillance blimp that hovered high over their base. The blimp used an expensive camera to monitor the region. It had effectively observed Taliban movements except in this one area where some trees masked Taliban comings and goings.

There were fourteen men on Michael's mission, including some EOD personnel. The group left from an intermediate structure called TI-2 (a tactical infrastructure) at 1600 hours (4:00 p.m.), walked to Girandai, immediately cut down several trees, and took a break.

Tactical infrastructures were often built when the distance from the FOB (forward operating base) to a town was considered an unsafe number of miles for a single journey. These TIs have few comforts; troops sleep on the ground, but the TI provides protection and brings the squad closer to a village for the next day's efforts.

That day, after cutting some more trees, the patrol planned to blow up any IEDs they found, which were many, and then wait in ambush for any inquisitive Taliban. After the trees were cut, the group approached Girandai; the squad leaders split their groups into two elements to search the area.

One element headed in one direction, and the other abbreviated squad in a different direction. DeMar led the second squad as its voluntary point man with a mine sweeper; some EOD personnel joined. Both groups moved with extreme caution, sweeping meticulously for IEDs.

Minutes later, the first squad heard an explosion and someone screaming for a medic. At first, the squad leader assumed that the

EOD team accompanying DeMar's group had purposefully exploded a discovered IED. He then realized he never heard the standard EOD one-minute warning before exploding ordnance.

The first squad leader reversed his steps and cautiously returned along his cleared route to find a collapsed, deafened, bleeding squad leader who had been peppered in the face by shrapnel.

Initially, he did not see DeMar but only a blast hole in front of the blinded squad leader and some of DeMar's equipment strewn about. After a few moments, he noticed DeMar up ahead on the other side of a partial wall in a collapsed condition. The experienced leader didn't rush forward, extremely cautious about the danger of additional IEDs. Carefully proceeding, he and the platoon medic avoided an IED in an archway on their way to help DeMar.

Feverishly, the squad leader and medic began treating DeMar, who was catastrophically wounded, unconscious, and yet had a slight pulse. While they made every effort to stabilize DeMar, a medevac helicopter was called, and security was established around them. EOD personnel found another three or four IEDs while the team was working on DeMar.

Because of the numerous IEDs, soldiers helped form a human wall to lift DeMar and carry him safely to the helicopter. DeMar and his wounded squad leader were quickly medevaced.

The two abbreviated squads were directed to stay in place for the remainder of the day as a show of force and return to base after dark, around 1:00 a.m. When they saw the Commanding Officer, First Sergeant, and Chaplain waiting for them, they understood that DeMar had not survived.

In the early morning hours on August 17, 2012, over 6,000 miles from Afghanistan, Michael and Lisa DeMarsico, the parents of Michael DeMarsico II, received the staggering news that their middle child had been killed in battle while voluntarily leading his squad through a dangerous minefield. He will be the first and only North Adams resident killed in action since the Vietnam War.

Immediately, with the help of the town's mayor, plans began to form to welcome Michael home. Municipal flags were ordered to be flown at half-mast, and the mayor met with Michael and Lisa to help arrange their son's homecoming. It was not the type of homecoming the parents had planned, but they wanted it to be memorable in honor of the person they called Mikey, his sisters and brother called Mike, and his Army friends knew only as DeMar.

His parents flew to Dover, Delaware, to welcome Michael back to America on August 18, 2012. Later, on August 29, his entire family traveled to Westover Air Reserve Base near Springfield, Massachusetts, to accompany Michael back home.

The seventy-mile trip from Springfield back home to North Adams witnessed thousands of people honoring Michael and his sacrifice. Individuals gathered in small or large groups along Routes 91 and 2, waving flags or holding signs with Michael's name on them, all paying respects to a young soldier who gave his all. Fire, police, and EMS units were stationed at every overpass along Route 91 and either saluted or presented flags.

At the roundabout circle in Greenfield, it was estimated that a thousand people were waving as the motorcade started over Route 2, bringing Michael home. People were holding flags through the small towns of Shelburne Falls and Charlemont, and bridge crews hung a flag from their cranes and doffed their hard hats. Massachusetts State Police and the Patriot Guard led the way.

When the Patriot Guard neared the top of Florida Mountain, they unfurled flags on their motorcycles as the procession headed down into North Adams, Michael's hometown. Drawing closer and approaching the well-known hairpin turn, the procession was greeted by standing salutes from the Clarksburg and Florida Fire Departments.

As the procession approached the center of town, a sea of red, white, and blue flags welcomed Michael home. A massive outpouring came from local citizens, including neighbors, friends, veterans, local college sports teams, companies, and people just wanting to offer their respect. The large crowds seemed to embrace the family and Michael.

Funeral services were held at the First Baptist Church, attended by Governor Deval Patrick, Senators John Kerry and Scott Brown, Major General Stephen Lanza, and state representatives. Both Pastor Dave Anderson and General Lanza spoke at the church. Pastor Dave spoke about Michael's early years growing up in North Adams and what he was like as a brother, with humorous tinges.

During the somber funeral service in a completely packed church, Michael's little sister, Leigha, was sitting close to the front of the church with her family when she noticed a bright blue dragonfly. It was flitting back and forth from Mike's coffin to each of her siblings. At one point, it landed on an elderly lady and startled her, causing Leigha to laugh. From then on, the dragonfly took on a special meaning for the family, symbolizing Michael's presence.

Michael was buried with full military honors at Southview Cemetery, with Pastor Dave officiating. Michael's mom, Lisa, was presented with the folded flag from his casket, and each of his sisters and brother was presented with a flag.

Michael's aunt had made a statement to the press shortly after hearing of his death. When interviewed by the *Transcript,* she said, "Michael was tragically killed doing what he loved . . . . He was a hero that was deployed to Afghanistan and who has now been deployed to heaven. He's just standing at a different gate. He was due to complete his deployment in December when he would also celebrate his 21st birthday" (8-18-2012).

Michael served approximately eighteen months in the US Army, five of them stationed in Afghanistan.

Michael's medals were given to the family by Casualty Assistance Officer Kenneth Shean. For heroic service, he received the Bronze Star Medal, a Purple Heart, the Army Commendation Medal, the Army Good Conduct Medal, the National Defense Medal, the Afghanistan Campaign Medal with service star, and the Global War on Terrorism Service Medal. He also earned the Combat Infantryman Badge, the Special Skills Badge for marksmanship, and the overseas service bar.

As deployments ended for Michael's platoon members, a number of them came directly to North Adams to offer their condolences and pay tribute to his family before returning to their own homes.

In memory of Michael, several ongoing programs, including the "Up Front for DeMar 5K Run and Walk" and the "Drive for DeMar" golf tournament, sponsor two annual $1,000 scholarships for Drury High School students. The North Adams Youth Football League now awards a trophy at the end of their season to the player "with the most heart." It's called the Michael DeMarsico Honor Trophy.

His sister Leigha is finishing her degree in criminal justice, emulating her older brother's desire to work in law enforcement. She wants to be a police officer working with youths.

This monument to Specialist Michael DeMarsico is outside the North Adams Armory. *[Mike Remillard Photos]*

When Michael was ready to deploy, he called his mom and said that he had named his parents as life insurance beneficiaries. He told them, "You've always lived in an apartment; if something happens to me, use the money to find your 'forever' house." Years after Michael's death, his parents were out for a drive near Southview Cemetery. They noticed a house being completed just across the street and up the hill from the cemetery, close to where Michael is buried. As they looked at it, a beautiful dragonfly landed on the house. With no hesitation, after seeing the dragonfly, they knew this was their forever home.

Michael's team members always said that he was very proud of his military service, and his leaders could see he cared about being the best soldier possible, desiring to learn more and be the best he could be. Ultimately, he volunteered for a most hazardous job to protect others, and there can be no greater tribute.

Williamstown, Massachusetts

Colonel Ephraim Williams Jr
Benefactor, Williamstown and Williams College
FRENCH AND INDIAN WAR

In 1755, a recently promoted forty-year-old Colonel was ordered to lead over a thousand men to intercept the invasion of hundreds of French and their Native allies near Lake George, New York. A colonial patriot from the Province of Massachusetts Bay, Ephraim Williams found himself in the center of a brutal and primitive struggle between England and France over the control of the Americas.

This courageous officer, though killed in the struggle, had provided in his will for a "free school" and a "township west of Fort Massachusetts." This township became Williamstown, and the "free school" became a men's college in 1793, now internationally known as Williams College, the second-oldest institution of higher education in Massachusetts.

THE WILLIAMS CLAN WAS one of the more prominent and influential families in western Massachusetts, part of a group referred to as the "River Gods," aristocrats who dominated river towns in the Connecticut Valley before the American Revolution. They owned vast farms, monopolized the shipbuilding industry, and supported local artisans. These families were wealthy when they arrived, building large manors, and only increased their wealth. The "Gods" ruled Hampshire County politics, involved themselves in militias, and even appointed the clergy.

Ephraim Williams Sr. was a well-to-do landowner and civic-minded citizen. His ancestors had emigrated from Norwich, England, in the early 1600s. Ephraim Sr. married Elizabeth Jackson in 1714, and they resided in Newton, Massachusetts. The couple had two sons, Ephraim Jr., born in 1714, and Thomas, born in 1718. Tragically, Elizabeth died shortly after Thomas's birth.

Their maternal grandparents, the Jacksons, raised the boys. Their grandfather was a property owner of considerable worth, who became involved in town affairs. Although early information on the boys is sketchy, it is presumed that they were educated in the village school. Thomas did attend Yale and became a physician.

After some time, Ephraim Sr. remarried and relocated to Stockbridge, Massachusetts, with his second wife and their three daughters and three sons. Ephraim and Thomas never returned to live with their father.

Little additional information exists about the younger Ephraim before 1740. He is known to have gone to sea for some time, possibly as a sailor but more likely working as a representative for his cousin's mercantile business trading in Europe. He traveled to England, Spain, and Holland, and one quote (attributed to a half-sister) noted that on these journeys he "acquired graceful manners and a considerable stock of useful knowledge."

Williams' library, along with his reading pursuits throughout life, gives some indication of his interests. His reading covered a broad range of topics from philosophy and spirituality to politics and history. His library included books on or by John Locke, Alexander Pope, and Daniel Defoe as well as the *Life of Oliver Cromwell, A Treatise of Military Discipline*, *Cato's Letters*, the *Universal History* series, the *Psalms of David*, and many political treatises.

Early records indicate that in the 1740s, at age 27, Williams moved to Stockbridge, Massachusetts, when he was asked to survey lands granted to his father. Williams also began purchasing land in the area. At this point he joined the local militia.

Ephraim Sr. was by now a selectman, a Justice of the Peace, and involved in a successful mission with the Housatonic tribe, selecting and supporting the missionary teacher who would instruct Housatonic children on agriculture and housekeeping. The mission also provided instruction in the Christian religion. The family of Ephraim Williams Sr. was one of four white families selected by the Governor as a model family for their mission work in Stockbridge.

As a backdrop to the younger Ephraim's adulthood, England and France were warring in Europe and the colonies, fighting over territories and trade issues. In the Americas, King George's War ran from 1744 to 1748. Governor Shirley of Massachusetts led the effort that besieged and captured the French fortress of Louisbourg on Cape Breton Island in Nova Scotia. It has been estimated that Massachusetts lost 8% of its adult male population during these years of conflict.

The French and Indian War followed from 1754 to 1763, during which each side continued to enlist local tribes as their proxies/allies, and much of the fighting and depredations were committed with their assistance. Combat was furtive, brutal, and usually deadly. Scalping was done by the French, English, and their allies, all rewarded by cash bounties paid in pounds authorized by state governors.

As the hostilities were increasing in the spring of 1744, during King George's War, England and its Iroquois allies fought against France and its indigenous allies. Colonel John Stoddard proposed to Governor Shirley a line of defensive forts from Connecticut to New York, each about five miles apart. The Governor concurred and obtained approval for their construction.

In 1745, Captain Ephraim Williams was appointed to a committee on the "defense of the province in time of war, both on the seacoast and inland frontiers thereof." Likely, Williams was promoted by Colonel Stoddard; he was put in charge of building and defending the recently erected line of blockhouses and forts, including Fort Massachusetts. He headquartered at Fort Shirley, named after the Governor and located in Heath, Massachusetts, about midway between the defensive structures.

Williams spent much of his time directing scouting parties that ranged between the forts, which were anywhere between four to six miles apart, directing his troops to seek out, intercept, and in many cases, provide advance warning for the settlers of any incursions by the French or their allies.

He spent considerable time ensuring his troops were supplied with ammunition, food, and equipment, even including snowshoes for the harsh winter conditions. Daily food requirements per militia-man were said to include a pound of bread, a pound of pork, and a gill of rum (approximately four ounces, or just under three shot glasses).

In the summer of 1746, when Williams was away from Fort Massachusetts, probably seeking supplies in Boston, it was besieged by a large force of French and Native allies, captured, and burnt to the ground. Governor Shirley quickly directed Williams to rebuild it, and some months later, with a work and guard force of 370 men, the job was complete.

During this period, Williams remained involved with the Stockbridge Indian Mission and provided for it in his first will, leaving money for "Christian Instruction." The same year (1748), Fort Massachusetts was again attacked, and Williams, while responding to the attack, was almost caught in an ambush by making an ill-advised foray outside the fort's walls.

Ephraim Williams survived, and he continued to buy and sell land in Stockbridge and Hatfield. In the process, he sold a slave, a "negro boy" named Prince, about age nine, in 1750. Prince may have been "acquired with" a house Williams had bought, writes Wyllis Wright in his biography of Williams, which contains a note in Williams' hand:

> *For and in consideration of the sum of two hundred and twenty-five pounds old tenor to me Ephraim Williams Jr., well & truly paid, by Williams Esq., of Hatfield, I do hereby assign, sell & convey a certain negro boy named Prince, age about 9 years a servant for life, to hold to him, his hiers [sic], against claims of any person whatsoever, as witnessed by my hand this 25th day of September Anno Domi. 1750. EPH WILLIAMS JR.*

During his lifetime, Williams enslaved at least four additional people: an adult named Moni, a boy named London, a girl named Cloe, and another boy named J. Romano, aged about sixteen years. Little else is known about these people. Until the mid-eighteenth century, slavery was common among well-to-do northerners, although not as central to their economy as in the southern states. By the 1780s, after several court decisions, the practice gradually disappeared in Massachusetts.

During the early 1750s, Williams petitioned the state and was granted 190 acres near Fort Massachusetts, provided he build a dam and a grist and sawmill (on the north branch of the "Hoosick River") within two years of being granted this petition and keep it in good repair for twenty years. For an unknown reason, he sold the acreage a short time later.

As hostilities seemed to ebb, he was last listed as the commander of Fort Massachusetts in late summer or early fall 1752 before leaving for Stockbridge. In Stockbridge, Williams became involved in town politics and was elected selectman. He remained very active in the Stockbridge Mission school, following in his father's footsteps.

During this time, Ephraim Williams continued to speculate on land, manage his father's estate, and care for his elderly parents. His father passed away in 1754. It is presumed that, while at Stockbridge, Williams got to know and become friends with Chief Hendricks and his family, who lived at the mission and were firmly aligned with the English government. (Chief Hendricks had been born in Westfield, Massachusetts, some forty miles away.)

In June 1753, Williams was commissioned by the governor as a major of the Southern Regiment of the Militia in the County of Hampshire—Foot Company of Stockbridge. As hostilities were again on the rise, Major Williams was reassigned to command Fort Massachusetts, adding to his area of responsibilities.

In early 1755, Governor Shirley began planning an expedition to overtake the French outpost at Crown Point and, with some success, hoped it would open the way to capturing the French headquarters in Montreal. Shirley had obtained authorization from England to raise two regiments. He offered Williams a commission in his regiment of lower rank (captain-lieutenant), yet it would come with the honor of leading one of his companies.

Williams did accept the commission. He was deeply convinced that Crown Point needed to be confronted. It was the source of the raiding parties of western Massachusetts. Governor Shirley directed Williams to begin recruiting immediately, with some guidelines, among them:

- recruits shall be between 18 and 40 years of age.
- only able-bodied men, free from bodily ails and of perfect limbs.
- no Roman Catholics, nor anyone under 5 feet 4 inches high without their shoes.
- enlistments shall be for 3, 5, or 7 years, with accompanying bounty.

Williams set about his duties earnestly, and then he received word that England would provide the officers for Governor Shirley. His appointment to Shirley's regiment had been therefore withdrawn. Sorely disappointed, he realized it would eventually work out to his benefit. General Johnson was assigned the formidable task of conquering Crown Point.

There were four planned military actions over a five-month period in 1755: the capturing of Fort Duquesne by General Braddock, Fort Niagara by Governor Shirley, Crown Point by General Johnson, and Fort Beausejour by Colonel Monckton. In the end, Braddock's efforts were a disaster and Shirley's efforts were abandoned. Fort Beausejour would be in fact captured and renamed Fort Cumberland; and though Crown Point, Williams' goal, was ultimately unsuccessful, it would do considerable damage to the French forces.

Williams received his rank of colonel in April 1755 when he joined General Johnson's regiment. In May, orders came from Governor Shirley to start the regiment on to Crown Point. The troops were ordered to muster at Westfield, Massachusetts, then march through Stockbridge, on to Kinderhook, New York, and meet in Albany. When they finally gathered, there was some confusion. Supplies were not ready, tents were missing, medicines not stocked, and battles (boats) built but not yet corked and tarred. The regiment was also in need of additional wagons and salt pork.

While waiting in Albany, Colonel Williams drew up a new and final will on July 22, 1755, leaving his property to his brothers, sisters, and other relatives, money to support his aging mother, payment of his debts and funeral costs, and detailed instructions on the distribution of personal items such as tankards, horses, silverware, books, pistols, parts of his library, and a large Bible.

Most importantly, upon his death, the interest from the sale of his remaining lands and any money from the sale of his bonds and notes shall be "appropriated toward the support and maintenance of a free school in the township of Fort Massachusetts (commonly called west [sic] township) forever . . . and provide also that the Governor & General Counsel give said township the name of Williamstown . . . ."

Thompson Chapel on the grounds of Williams College, established in his name in 1793. *[Mike Remillard Photos]*

Not long after he completed his will, the commissary filled the supply shortages, and they were off. The regiments began wending their way along the fifty-mile trek from Albany to Fort Lyman, often needing to cut a swath through the woods, ensuring slow progress. In late July, the first rumors began spreading of General Braddock's devastating defeat; this unnerved the regiment.

Williams' men proceeded in increments, cutting roads, hauling supplies, carting small pieces of artillery and portaging battles, as well as setting up small base camps; it was a slow and highly fatiguing process. Everyone was on heightened alert after hearing about Braddock's rout. Security and safety were an ongoing concern; one militia man's thigh was broken when a wagon ran over him, while another had his arm amputated by an accidental discharge.

Once reaching Fort Lyman (later to be renamed Fort Edwards), they learned it would become the regiment's supply base. The troops continued to inch their way northward toward what would become Lake George, about twenty miles distant. Hundreds of soldiers worked on cutting a road while alarming news continued to filter in from their Native scouts that French troops were reinforcing Crown Point.

In late August 1755, General Johnson, his 1500 men, and half his artillery reached the southern shore of Lac du Saint Sacrement, and he renamed it "Lake George" in honor of his sovereign, George II. Johnson's army was quickly reinforced and grew to approximately 2850 soldiers. Ephraim Williams and his regiment were with him, though the group was still fifty miles from their destination of Crown Point. The troops started constructing a base camp that someday would become Fort William Henry. Colonel Williams was designated to oversee its construction.

His orders were quickly changed when reports were received of nearby marauding enemy bands. The French, realizing General Johnson was on his way to capture Crown Point, had dispatched a wily, accomplished frontier fighter named Baron Dieskau with 1500 soldiers composed of French regulars, Canadian militia, and hundreds of Native tribesmen, including Iroquois and Mohawk, to deal with Johnson's troops. Rather than attack Fort Edwards and its formidable cannons, the French and their Native allies decided to lay in ambush not far outside the fort between the two forces (Johnson at the lake and the other at the fort).

North of the fort near Lake George, Colonel Williams was put in charge of leading a thousand soldiers and two hundred Native allies, almost half of General Johnson's forces, back to Fort Lyman to protect the critical supply base. Williams' friend, Chief Hendricks, who had more than forty years of scouting, led the advance guard with his Mohawks, and Williams was to be right behind him, leading the main body.

As the long column approached a narrow ravine, a spot most perfect for an ambush, Chief Hendricks is alleged to have said, "I smell Indians," and at the same time, a small herd of spooked deer ran by. Neither served as sufficient warning.

The column continued. The enemy was quietly set up atop both banks flanking Williams' troops, prepared to fire directly into the unsuspecting, approaching men. Others were positioned to fire into the rear of the assembled troops. The hundreds of ambushers placed on the high ground also had the advantage of being able to fire into the open from behind trees, brush, and rocks. Baron Dieskau had placed his men well.

The ambush was a scene of absolute chaos, and the surprise engagement became known as "The Bloody Morning Scout." Hundreds of musket balls were simultaneously launched at the English force from above. Black powder smoke filled the air as the enemy reloaded their Charleville muskets for at least one more shot before hurling themselves at the troops below. (The butt of the Charleville is also known as *patte de vache*, French for "cow's paw"; its shape makes for an excellent club.)

Once a musket was discharged a second time, it was often repurposed as a lance with a 20-inch bayonet. In these primitive, vicious, hand-to-hand fights, bayonet wounds accounted for a third of the casualties. Chief Hendricks' horse was shot out from under him, and he was bayoneted twice by one of the tribesmen fighting alongside the French.

During the battle's initial intensity, Colonel Williams is said to have raced up the slope, jumped atop a large rock, and attempted to rally his men. He was shot in the head even before he had a chance to fire his weapon. Members of his regiment hid his body under some brush to avoid it being mutilated, and they retreated with the rear guard.

With such a surprise and massive torrent of bullets, some of the new recruits quickly retreated the four miles back to Johnson's base camp, and Dieskau's forces were right behind them. The British set up a rearguard to protect those that were fleeing.

General Johnson heard the distant firing from his position at the base camp. He had his men begin fortifying it, and they were well prepared for the coming assault. French forces attacked the camp repeatedly, and although Johnson was wounded, the French had significant losses; at one point, the English left their barricade and charged the French, inflicting more losses. Dieskau was wounded and captured, and the French retreated. The English were also able to ambush the French baggage train, inflicting another 300 casualties.

Ironically, Dieskau, who was wounded several times, was probably treated by Dr. Thomas Williams, Ephraim's younger brother and a survivor of the battle. This Yale-educated, accomplished regimental surgeon had been appointed by Governor Shirley. (Their cousin, William Williams, another survivor, would be a future signer of the Declaration of Independence from Connecticut in 1776.)

Several days later, a burial party of 400 soldiers buried 136 troopers from the morning ambush. Colonel Williams had been buried under a pine tree by the side of the road. His body was found undisturbed where it had been left.

"His sword and watch, with other personal effects, were taken back to the camp and entrusted to his brother Thomas, whose descendants, one hundred years after the battle, presented them to the college that bears his name," writes Wyllis E. Wright in *Colonel Ephraim Williams: A Documentary Life.*

Colonel Williams was criticized for not having flanking guards out as his column searched for the enemy, a common practice for an experienced leader. Others presume he was using the Mohawks as his reconnaissance group, intending that they would draw fire and

Ephraim Williams' watch can be seen today at the Special Collections, Williams Libraries, Williams College. *[Mike Remillard Photos]*

then his troops would engage the exposed enemy. No one would ever know his motive. In the end, Dieskau had all the advantages.

Williams' battle chest with his personal belongings was inventoried and returned to his family. Its contents were listed as follows:

- One sword belt
- One Psalm book and One Testament
- One pair of breeches with silver buttons
- One pair of flannel holsters
- One pair of Indian shoes, beaded
- Two pairs of leather stockings
- Razors
- Two red worsted caps
- 5 checked shirts
- One broadcloth coat with yellow buttons
- 1 French bearskin coat, white buttons metal
- One wig box & comb
- Two woolen vests
- Two pairs of striped linen trousers

In 1854, at the original site of the Colonel's death at Lake George, Williams College alumni erected a white marble obelisk that still can be seen today atop the boulder where Colonel Williams was killed. The alumni enclosed the site with a black wrought iron fence. In 1931, workmen digging near Williams' gravesite uncovered the remains of four soldiers from that long-ago battle; all the skulls were identified as of "European" descent and having died of head wounds. Once they were identified, the remains were buried near the battlefield.

This skirmish opened the Battle of Lake George. Despite the early ambush and death of Colonel Ephraim Williams, Lake George was an early victory for American provincial forces over the French and their Native allies. General Johnson was hailed a hero and made a baron. Moreover, it was considered one of the critical points in American history of the struggle for territorial control of North America by the French and English. General Johnson is considered to have won the overall battle.

It is said that in 1834, Colonel Williams' nephew, Dr. William H. Williams, returned to the site of the battle, located his uncle's grave, removed his skull, and carried it away. Eventually, in 1920, Williams College had Colonel Williams' entire remains transferred to a vault at the Thompson Memorial Chapel on campus.

Undated pen and ink sketch of Ephraim Williams, Jr. *[Under Public Domain]*

During Williams' lifetime, descriptions of him seemed to describe a caring individual, affable and unpretentious. He was civic-minded and involved in his community. Williams helped his soldiers and their families with their needs and made sure they were paid on time. He spoke up for his men and their families when their enlistment money was delayed, supported cash allowances for bringing their own firearms,

A monument to Ephraim Williams, Jr., holds a place of prominence in Thompson Chapel. Downstairs lies the vault containing his remains, dated 1755 and marked with his initials. *[Mike Remillard Photos]*

and made sure they received knapsacks and blankets. Williams had a reputation for being thoughtful of the men who served with him. He effectively led his soldiers while caring for them, and they knew he would also share any risks he might ask them to take.

Respected and honorable, he spent considerable time and money supporting the Stockbridge Mission and its efforts in assisting indigenous people. He did own enslaved people either as part of the property he bought or inherited; they were not numerous but nonetheless, he owned slaves.

Williams remained single his entire life. It is said that he fell in love with Sarah Williams, a nineteen-year-old much younger than himself, but she was already interested in another person.

The only description or epitaph we have of Ephraim Williams is from a recorded statement in 1802 by Ebenezer Fitch, the first president of Williams College, whose term ran from 1793 to 1815. Fitch was living in Williamstown some forty-seven years after Colonel Williams' death. He had spoken with some elderly survivors of the old garrison of Fort Massachusetts who knew Colonel Williams; they had described him thus:

"In his person, he was large and fleshly . . . , He had a taste for books and often lamented his want of a liberal education. His address was easy, and his manners pleasing and conciliating. . . . Affable and facetious, he could make himself agreeable in all companies; and was very generally esteemed, respected, and beloved. His kind and obliging deportment, his generosity, and condescension [as in, not putting on airs], greatly endeared him to his soldiers. He was uncommonly beloved by them when he lived and lamented when died. . . . His politeness and address procured him a greater influence at General Court than any other person at the day possessed. He was attentive and polite to all classes of men, but especially to gentlemen of dignified characters; and sought the company and conversation of men of letters" (*Colonel Ephraim Williams*, Wright, p. 150).

One wonders how Williams would have had the desire or foresight to create a highly competitive school in the primitive area of West Hoosac over two hundred years ago. Presumably, his extensive travels throughout the Berkshires, defending the area and its residents and witnessing their hard lives in this isolated area, might have compelled him to philanthropy. The will that he composed, essentially on his way to sure death, was consequential in its call for support and maintenance of a free school in West Township.

It is remarkable to think that today's thriving community of Williamstown, with Williams College at its heart, is all the result of one man's vision and sacrifice so many years ago.

Ishmael Titus
Enslaved Militiaman
FRENCH AND INDIAN WAR
REVOLUTIONARY WAR

He never knew his parents, never knew the date of his birth, and only knew he was born somewhere in Amelia County, Virginia. Different people had owned him, and yet Ishmael Titus participated in some of the landmark battles of the eighteenth century, initially as a slave, then as a militiaman, and he helped establish America as a country.

Ishmael became a Williamstown resident for almost thirty years. He is presumed to lie there now.

IT IS ESTIMATED THAT at least 5000 free and enslaved African Americans served in the Continental Army, Navy, and militias during the Revolutionary War. As laborers, body servants, and front-line troops with no thought of segregation, they served side-by-side with their counterparts. Ironically, perhaps, they were fighting for America's freedom (not their own) from Great Britain.

Over 20,000 African Americans fought alongside the British, encouraged by their commander, who offered freedom to slaves who would fight for the king. The British also used runaway slaves as guides, spies, and laborers.

Ishmael Titus was born enslaved on the estate of Harry Bluford around 1746 in Amelia County, Virginia. Little is known of his parents or youth, and it is generally assumed he was not offered education. During the French and Indian War, his owner was employed by the

British commissary to supply British troops during Major General Braddock's ill-fated attempt to capture French Fort Duquesne. This massive endeavor by 2100 soldiers led to an arduous trek through one hundred and ten miles of wilderness.

Braddock's emissaries had secured one hundred and fifty wagons and hundreds of pack horses to carry provisions for his large army. The wagons were well-engineered, twelve-foot-long Conestogas, each with a team of four horses capable of transporting a considerable quantity of supplies.

Bluford, Ishmael's owner, seizing this opportunity, provided the British with a number of Conestogas to haul their supplies, and he also sent young Ishmael, at age nine or ten, to work on one of the heavily loaded wagons. Ishmael was tasked to ride one of the team horses.

The expedition traveled slowly as Braddock wanted to create a passable road for future use, especially to resupply his troops, and the dense wilderness was difficult to clear. The vanguard made slow, tiring progress with soldiers in the lead felling giant trees, moving earth, and cutting passages through hills.

The expedition had begun on May 29, 1755, and by July 8, over forty days later, they were still some ten miles south of Fort Duquesne. The French garrison at Fort Duquesne, though undermanned, was by now well aware of Braddock's presence.

At the rear of the detachment, sometimes over a mile away from the vanguard, Ishmael toiled with the supply group. When stopping at night, the work at encampments was continuous for them, loading and unloading wagons, setting up campsites, chopping wood, building fires, herding cattle, slaughtering, and cooking beef for the troops; as a young boy, Ishmael Titus was everyone's laborer.

The following day, he was involved in breaking down the campsite, folding the tents, and repacking the wagons. Ishmael knew little rest.

This established routine dramatically changed on the morning of July 9, 1755, when the British unexpectedly engaged the French and their Native allies and fought a running skirmish on narrow wilderness trails. The French and their allies, who fought from concealment, baffled the British troops who stood fighting in column order, and began decimating the British. Although the British greatly outnumbered the French, they fell back in disarray, and an orderly retreat became a rout.

Several miles back of the vanguard, Ishmael could hear the heated exchange of gunfire as he was helping to dismantle the previous night's tents and put out campfires preparing for the day's journey. The gunpowder smoke and its acrid smell were starting to drift his way. Suddenly he could hear men crying out, then there was a rush of red-coated soldiers, unarmed, frightened, and wounded, running by his wagon, and he did not know what to do. In his later years, he remembers the red coats of the British soldiers being the "color with blood," unfortunately, true in this instance.

The stream of troops grew; their red coats seemed so bright, many stumbling, some falling, unable to get up. His wagon wasn't loaded, and the horse team had not yet been harnessed to it.

Ishmael considered harnessing the team, but there were too many fleeing soldiers, and the trail was too narrow. Finally, an older wagoner who had befriended him at evening camp told him to run; the French and their Native allies were right behind them, killing and scalping everyone they found.

Ishmael quickly untethered the horse he had been riding for the past few weeks and escaped, just as he saw the enemy approaching the woods. The French and their allies did not pursue the fleeing redcoats for any great distance; instead, they resorted to looting abandoned supply wagons and scalping the dead and dying.

Thus began Ishmael's long trip back to Virginia. After weeks of travel with others, and days by himself, Ishmael returned to his owner, Bluford.

Those wagons and cannons that escaped capture were ultimately destroyed by the British so as not to fall into enemy hands. Of Braddock's 2100 soldiers, almost half were killed or wounded. Braddock himself was shot off his horse, mortally wounded, and evacuated from the scene by his junior officer, twenty-three-year-old George Washington. At the time of his death, Braddock was buried in the center of the trail, and his grave was run over by wagons so as not to be discovered by the enemy.

Several years after returning to Virginia, when Ishmael was around thirteen, he was sold to John and Dick Muir (Mara), who resided in North Carolina. They owned an iron company, and its water-powered blast furnace forged iron pieces. It is presumed that Ishmael worked here, either in the forge area or doing arduous pick and shovel work, digging and transporting iron ore; in either case, it was difficult work for a teenager. He was with the Muirs for some unrecorded time and then sold to Lawrence Ross.

Ross lived in North Carolina. As the American Revolution approached, the military required more men for service. Ross was to be drafted around 1779 into the North Carolina militia to serve a one-year enlistment. Permitted to send a substitute, he sent Ishmael in his place, who presumedly had little choice in the matter. As reported in the book *Eminent Charlotteans* by Scott Syfert, Ross promised Ishmael his freedom in exchange for this endeavor, but it was likely a verbal contract as years later, Ishmael was unable to produce documentation of his militia service.

ISHMAEL TITUS

1745/6 – 1855

As imagined by Makayla Binter in 2021

Born somewhere in southern Virginia around 1745 or 1746, Ishmael Titus was one of many enslaved men who served on both sides of the American Revolution.

Most of what we know about Titus's service in the American Revolution comes from the pension request he submitted in 1832. According to the records, when enslaver Lawrence Ross was drafted, he sent Titus in his stead. Titus served in numerous battles of the Southern Campaign, particularly the pivotal battle of Kings Mountain which helped turn the tide of the war. Titus saw combat and may have also engaged in much of the hard, manual labor associated with the building and transporting of large military encampments.

Unfortunately, like so many enslaved people who served, Titus's pension request was denied. It took until 2013 before Titus's service was officially acknowledged by the state of North Carolina. Today, his name appears on a plaque near the Harvey B. Gantt Center for African-American Arts and Culture honoring and celebrating the service of African Americans in the American Revolution.

Ishmael Titus, as imagined by artist Makayla Binter. The image hangs in the Charlotte Museum of History. *[Makayla Binter, Courtesy of the Charlotte Museum of History]*

In the spring of 1779, Ishmael Titus entered military service. He was stationed at a fort in South Carolina for a year and was involved in skirmishes with Loyalists and Native tribes. When his assignment expired, he decided to enlist in the Army.

There is some conjecture as to why he enlisted. Did he doubt his former owner would keep his word about granting him his freedom? Did he prefer Army life to being enslaved (one would expect so), or maybe he had nowhere else to go? At any rate, he enlisted in the company of Captain John Cleveland around June of 1780. The captain was the son of Colonel Benjamin Cleveland, a militia commander.

Under Captain Cleveland, soldier Ishmael's role became more official. The company practiced frequent drills and tactical training and learned to shoot from cover and concealment. Ishmael received a captured British "Brown Bess" musket and learned to use a bayonet. Under the Captain's tutelage, he also began practicing rapid-firing his single-shot muzzleloader until he could get off four accurate shots in a minute.

The company commissary helped outfit him with a worn knapsack, powder horn, bullets, canteen, haversack, and a blanket. A grizzled older militiaman had given him an old tomahawk and a sheath knife. Over time, he replaced his worn shoes and secured a warmer shirt and a wool coat.

In mid-August 1780, Ishmael Titus, along with Captain Cleveland's company, conducted a diversionary raid south of Camden Court House in South Carolina. They were part of a larger Continental Army attack led by General Gates north of the courthouse.

The battle was a disaster for the colonist forces. The smaller group of British troops routed the colonists in an overwhelming victory that strengthened Britain's control of the Carolinas. It took a little over one hour for the British to hand the colonists a humiliating defeat.

The colonists were not sufficiently trained or organized, and they quickly retreated from the battle-hardened, bayonet-led British formations. Their lines collapsed; they threw down their arms and fled the area. Ishmael and his company joined them in fleeing.

The British cavalry relentlessly pursued the broken militia for over twenty miles, killing and wounding many of them. The roadway leaving the battle site was strewn with broken carriages, abandoned cannons, screaming horses, and dying men. General Gates himself was said to have disgracefully galloped over sixty miles to escape the British. Half of Gates's colonial forces were killed, wounded, or captured, and he never commanded troops again.

Ishmael and his fellow soldiers retreated to Charlotte, North Carolina, and even further north as the British continued to follow them. Eventually, they recovered and refitted in a wooded area somewhere south of Virginia, but not for long.

Throughout the bitter fighting in the Carolinas, the continual battles, and the many defeats at the hands of the British, it is noteworthy that almost all those of African descent remained loyal to their militias. The opportunities to change sides were plentiful, could have been tempting, and likely seemed to have been the wiser course of action.

The colonist defeat at Camden Court House emboldened the Loyalists, who began controlling the countryside, rooting out and decimating local Patriot militias. (Note: The words *Tory* and *Loyalist* are synonymous. Both opposed the *Patriots* and attempted to prevent the colonies' independence from the British Empire.)

Realizing the jeopardy they faced, groups of frontiersmen from west of the Appalachian Mountains known as "Overmountain Men" banded together with southern militias to fight off the Loyalists.

These two large Patriot militia groups converged less than sixty days after the Battle at Camden Court House, on October 7, 1780, at a place called Kings Mountain, North Carolina.

More than 1000 Loyalist forces, mainly from North and South Carolina, realized that a battle was imminent and occupied the top of Kings Mountain. After Camden, the overly confident Loyalist militia, led by an experienced, professional British soldier named Major Patrick Ferguson, declared, "All the rebels in hell could not push" him off the top of the mountain.

After the Camden debacle, Ishmael had joined Colonel Cleveland's battle-weary Wilkes County militia from North Carolina to face the daunting task of dislodging Ferguson's Loyalist militia from the heights of Kings Mountain. In the process of being refitted, Ishmael had obtained a Kentucky long rifle which allowed him to shoot accurately at much further distances than his old British "Brown Bess."

Historian Bobby Moss notes in *Eminent Charlotteans* that "there were at least five African Americans who fought at Kings Mountain, including Ishmael Titus" (p. 90). Around 3:00 p.m. on October 7, 1781, Ishamel and Colonel Cleveland's militia joined other militias, and alongside the Overmountain Men, they silently clambered up King's Mountain surrounding the crest. Together they killed the outlying sentries and continued their advance. Nearing the top, they were discovered, and sporadic firing broke out, turning into a hailstorm of bullets. Yet the Patriots continued upward, scrambling from tree to tree. Initially, the Tories fired over their heads, but then quickly adjusted their fire, and Patriots began to fall.

The experienced frontiersmen and vetted militiamen used the concealment and made their shots count. Colonel Cleveland's group, including Ishmael, advanced directly into the gunfire. Cleveland himself had several horses shot out from under him.

Major Ferguson, the Tory leader, led a number of mounted, downslope bayonet charges that dislodged the Patriots and sent them tumbling down the mountainside. For a few moments, it looked like his efforts would turn the tide of the battle in favor of the Loyalists.

The Patriots regrouped and continued to advance on all sides. At one point, a frontiersman sharpshooter shot Major Ferguson from his horse, and he was hit several times as he fell and died. The volley of shots directed at him also killed several of his officers. Fighting became more intense as the militia and frontiersmen closed with the Tories. Tomahawks met bayonets as the groups clashed in hand-to-hand combat.

Not long afterward, firing began to slow, and pockets of Loyalists began surrendering, throwing down their rifles, although some of the angry Overmountain Men continued fighting and killing.

In a matter of sixty-five minutes, over 500 men had been killed or wounded on the 600-yard crest, all Americans except for one British man, Major Ferguson. Four hundred Tories were killed or wounded, and fewer than one hundred Patriots. Over 700 Tories became prisoners, and the Overmountain Men hanged nine of the more notorious ones. Surprisingly, other than minor cuts and bullet holes in his clothing, Ishmael survived unscathed.

The defeat at Kings Mountain caused the British General Cornwallis to leave Charlotte and retreat south, giving up any designs on proceeding to North Carolina and Virginia. This created a panic among the local Loyalist population that up until this point had become quite brazen; Cornwallis' abandonment dampened the attraction to their cause. It was a link in a long series of events that would eventually result in America's freedom.

The battle at Kings Mountain has been described as the largest "all-American fight." Years later, Teddy Roosevelt described it as a "turning point in the Revolution."

The war dragged on in the south, and at one point, General Washington announced that General Nathanael Greene would be taking command of the southern department of the colonial forces. Greene found his southern army in poor condition, lacking supplies as essential as blankets, clothing, footwear, and bullets.

Ishmael's wartime service wasn't over; he was still serving faithfully in the Wilkes County militia under Captain Beverly. He was hungry and shoeless. The ill-equipped troops still faced British troops actively trying to destroy them. With General Greene's quartermaster logistical experience, he worked hard to overcome the shortages and evade the British while preparing for the next battle and choosing a site that would advantage his side.

In March 1781, after eluding the British General Cornwallis for days, General Greene made a stand at Guilford Courthouse in North Carolina. Cornwallis was determined to destroy the primary Patriot force in the south. Cornwallis had 2100 battle-hardened veterans, and Greene had 4500 troops, many of whom were militia. Ishmael was part of this group.

The Courthouse sat on a wooded, brushy site and served Greene's men well, disrupting cavalry and bayonet charges. The first contact was made early on the morning of March 15, 1781, and commenced with artillery bombardments by both sides. Then the battle devolved into small firefights by lines of men, with Ishmael in the thick of the battle. The British continued to advance, pushing the Patriots back, and finally, General Greene, after a ninety-minute battle, had his troops conduct an orderly retreat.

Despite having fewer troops, the British occupied the battlefield at the end of the day, although in "winning," the British lost over 25% of their total force, the Patriots less than 10%. In Britain, a parliamentarian declared that "another such victory would ruin the British Army."

The battle was considered to have stopped British momentum in the south and changed the direction of the war for the Patriots, generally freeing North Carolina of redcoats.

For Ishmael, this was his last battle. He recalled feeling as if "I marched through the (entire) State of North Carolina" when it was all over. He went on to say, as quoted in *Eminent Charlotteans*, "'I was discharged on the Holston River at the Log Court House.' Although he 'cannot recollect the name of the place,' he knew it was at the close of the war" (Page 90).

Even after his discharge, the excitement wasn't over. On the way home, traveling over the Allegheny Mountains, he and his party of ten, including his commanding officer Colonel Cleveland, were captured and tied up by a band of Tories led by Captain Riddle. The Tories were collecting prisoners for reward by the British.

Ishmael was forced to tend and feed horses. While foraging for the horses, he was found by two rescue parties led by Colonel Cleveland's son. The rescue party took the kidnappers under fire and scattered them, rescuing everyone. Weeks later, the kidnappers were captured, and Ishmael saw nine of the marauding Tories hanging outside the local Courthouse.

After his discharge, beginning sometime in the 1780s through the early 1800s, Ishmael moved from place to place frequently. There is a presumption that he was concerned about his status as a "freedman." Later in life, he is quoted as saying he "ran away from his master." One wonders if the deal made for his freedom, enlisting in place of Lawrence Ross, was ever honored. Was he looking over his shoulder, worrying about slave traders?

Records indicate that he repeatedly moved throughout New York State. Ishmael Titus married Thankful Shepard on November 13, 1805, and later a second wife, Lucy Rogers, on September 5, 1812, at age 66. Ishmael and Lucy moved to Bennington, ultimately residing in Pownal, Vermont.

The couple moved to Williamstown, Massachusetts, sometime after 1820. Records indicate they lived in a little hamlet north of town known as White Oaks, in those times occupied mainly by a "colored population," most of whom came from New York State when slavery was abolished there. Ishmael and his wife lived in a house near Broad Brook (a good trout stream), and their son Harvey lived in a house north of the couple.

For most of his life, Ishmael struggled with financial stability as an emancipated yet uneducated old man with few skills; he did develop a relationship with a farmer in Bennington who bought some of his handmade wares, but Ishmael had a limited existence, and later in life he began to go blind. At one point, he started a petition collecting signatures and donations to remove a mortgage from his home.

The years had exacted a toll on Ishmael Titus. He was thin, frail, and was said to have a large "wen" on his neck. His memory of long-ago battles in the hills and backwoods of the Carolinas was sketchy.

On October 10, 1832, at age eighty-six, Ishmael appeared in court to apply for a Revolutionary War pension from the federal government. The Pension Act provides every surviving soldier who served at least two years in service to his country a pension with full pay. It was a benefit he sorely needed.

While acknowledging to the court that he was at the battles of Kings Mountain, Guilford, and Camden, it was sometimes only after consulting with history books and reading descriptions of the commanders, locations/terrain, and enemy combatants. He could not produce any documents of enlistment or discharge since militia records were almost non-existent.

Four persons did vouch for his integrity, but he lacked proof of his age and any concrete evidence of his service. The court consequently denied his petition for a pension.

Ishmael continued to live in Williamstown for well over twenty-five years, enjoying his children and grandchildren and entertaining them with stories from his early travels throughout the Carolinas. He died in Williamstown on January 27, 1855, at the impressive age of 110. His death was carried in papers across the eastern states, many of which ran as the heading, "Death of the Last Survivor of Braddock's Defeat."

Ishmael Titus is presumed to be buried in Williamstown, although the location of his grave is unknown.

Not long after his death, a certain famous author wrote a tribute to the Patriots of the Revolution. It seems fitting to consider these words in the context of Ishmael Titus and his sacrifice, denied as it was by the country whose independence he helped secure:

> *In considering the services of the Colored Patriots of the Revolution, we are to reflect on them as far more magnanimous because rendered to a nation which did not acknowledge them as citizens and equals, and in whose interests and prosperity they had less at stake. It was not for their own land they fought, not even for a land which had adopted them, but for a land that had enslaved them, and whose laws, even in freedom, oftener oppressed than protected. Bravery, under such circumstances, has a peculiar beauty and merit.*
>
> ***—Harriet Beecher Stowe, October 1855***

Colonel Charles W. Whittlesey
Medal of Honor
World War I

In World War I, a Berkshire man successfully won a desperate fight in the Argonne Forest. He led the survivors of two decimated battalions that had been surrounded, cut off, and viciously assaulted by German infantry for five days. He disregarded a demand to surrender, and fought on with no food, little ammunition, many gravely wounded, and scant hope of survival.

The Lost Battalion, as Charles Whittlesey's men became known, was a misnomer, an accidental double entendre. A United Press journalist intended to indicate the group was facing complete annihilation, not that they were physically lost. In actuality, they had reached their military objective, yet the area was so dense their command could not find them.

IN FLORENCE, WISCONSIN, CHARLES White Whittlesey was born on January 20, 1884, to Frank R. Whittlesey and Anne Gibbs-Whittlesey, who married in 1881. The family's heritage traced itself to Puritan colonists. Charles was the second eldest of six boys, four of whom reached adulthood. His one sister, Annie, died of "black diphtheria" at a young age.

Born in Danbury, Connecticut, Frank had moved west and worked in the lumber, mining, and real estate industries, eventually settling in Florence, Wisconsin, in 1890. He received an official appointment by President Benjamin Harrison to be the city's postmaster. It was there that Charles attended the Old First Ward grammar school.

In 1893, the family moved to Pittsfield, Massachusetts, where Frank worked as a purchasing agent and, later on, as a production manager for the General Electric Company. The family resided at 38 Pomeroy Avenue, and Charles attended and graduated from Center Grammar School in 1897. Its class motto was "A Thing Well Done, is Twice Done."

Charles graduated from Pittsfield High School in 1901 and was accepted at the prestigious Williams College, twenty miles from home. By all accounts, the tall, slender, bespectacled youth, a modest and sensitive man, was well known, liked, and displayed early signs of leadership and caring for others. As a member of the Class of 1905, he quickly became involved in campus life.

Charles liked to write, contributing to campus newspapers, and becoming the editor of the yearbook, *The Gulielmensian*, nicknamed "the Gul." He was a member of the Delta Psi fraternity and was elected to the Gargoyle Society, which defined itself as intending to "promote students' moral, intellectual, physical and social growth."

The serious side of the young Charles Whittlesey. *[Courtesy of Special Collections, Williams Libraries, Williams College]*

He was also the Editor of *Williams Literary Monthly* and the *Williams Record* and contributed to both frequently. Prolific at writing, Charles received a $15 prize for the most significant number of articles submitted in 1902. He wrote an essay on the "Literary Enterprises" of the Class of 1905 for the yearbook, joined the Philologian Society, and also served as toastmaster at various suppers.

Classmates had two nicknames for him: "Chick" for sociability and "Count" for his reserved nature. His class also voted him the third brightest man in 1905. Other than writing, Charles loved college football games.

Upon graduation, he was accepted at Harvard Law and graduated in 1908. One of his better-known classmates was Theodore Roosevelt, Jr. After graduation, Charles joined a New York City law firm, working there until 1911 when he joined a friend in private practice, eventually leaving in 1917.

As a backdrop to this period, the war had raged in Europe since 1914. The United States was determined to maintain a position of neutrality and continued to engage in commerce with all European countries. Charles supported President Wilson's position and thought war to be immoral.

As the possibility of war increased, the United States began converting its pioneer army to a more professional armed force. Realizing the eventuality of war, the government created the elite Plattsburgh Officers Training Camp to expand its forces and train young Ivy League–educated men who would lead others into battle. In August 1916, Charles and his brother Elisha attended and graduated from Plattsburgh along with Teddy Roosevelt, Jr.

Germany's sinking of the ocean liner *Lusitania* in 1915 with hundreds of innocent American passengers on board changed the public (and Charles's) opinion against Germany. With the continued sinking of more merchant ships, the United States declared war on Germany in April 1917.

Commissioned a Captain in the officer's reserve corps and called to active duty on August 8, 1917, Charles reported to Camp Upton in Long Island, where the 77th Division assembled and trained. The Selective Service Act was authorized in 1917, conscription began, and it had a high success rate, primarily due to immigrant patriotism. The United States was unprepared for war, having only a standing army of 127,500 soldiers. By war's end, over four million men had served.

Charles was assigned to the 308th Infantry Regiment (of the 77th Division) and initially given a staff position with Headquarters Company. The division quickly became known as the "Melting Pot" or "Times Square Division" and wore Statue of Liberty shoulder patches. The 77th Division was an amalgamation of poor immigrant draftees, many living in poverty on the Lower East Side of New York City. It was said that the group spoke 42 different languages, not including English. (Songwriter Irving Berlin served as social director for the 77th Division and arranged for movie stars to entertain the troops, among them Eddie Cantor and Sophie Tucker.)

America needed all available manpower, so the immigrant draftees were readily accepted into the army, but not without suspicion. Many considered them not to be 100% American (native-born). The question persisted whether they could be trusted as immigrants or would they take the side of the European countries.

All three of the Whittlesey sons were at this point serving in the military. Captain Charles W. Whittlesey was with the infantry in New York; Lieutenant Melzar N. Whittlesey was stationed at Fort Strong, Boston; and Elisha Whittlesey had served in France with Camion Ammunition Services. (In 1911, Russell, a younger brother, had died while at school of pneumonia and a heart condition. He had been twenty-three years old.)

After months of training, Charles's regiment, the 308th Infantry, sailed to France on April 6, 1918, as part of the American Expeditionary Force. The 77th was one of the first units sent overseas. The division fielded 27,000 men in four infantry regiments (305, 306, 307, and 308), two machine gun brigades, and field artillery.

The month-long trip ended in Picardy, France; Charles's unit was initially issued Enfield rifles and joined in defensive maneuvers with the British and French. In May 1918, the regiment moved to Flanders, Belgium, was issued Springfield rifles, and began patrolling and experiencing trench warfare.

Initially assigned to be the 308th's Operations Officer, Charles was promoted to Major and given command of the 1st Battalion of the 308th Regiment. Dispatches already noted his "coolness under fire."

In September, his unit was charged with assaulting the Argonne

Charles, shortly after assuming command. The photograph is autographed to his friend "William." *[Courtesy of Special Collections, Williams Libraries, Williams College]*

Forest. The area had been occupied and fortified by the Germans for four years. It had two lines of trenches supporting artillery and mortars and was considered by some to be an almost impenetrable area.

A highly dense section of woods, characterized by foggy and muddy trenches, made it a challenge to communicate with other units. The thickness of the forest canopy limited their communications to runners or carrier pigeons.

In late September, Charles's 1st Battalion pushed through to L'Homme Mort's cemetery and was cut off from any support. They endured three days of hard fighting before help reached the unit. His men had been told the fighting would be brief and thus had left behind shelter halves, raincoats, overcoats, and blankets.

The half-strength 1st and 2nd Battalions were no sooner rescued than told their next operation was to drive ahead at all costs, even without flank protection, and capture a hill near La Paulette. They were to move on and dislodge the Germans from Charlevaux Mill.

Told to "advance regardless of your losses," Charles acutely realized the dangers of the mission. He is quoted as saying, "All right, I'll attack, but whether you hear from me again, I don't know." Anticipating a brief encounter, the unit drew only one day's rations with the plan of quickly securing the hilltop. It was a bitterly cold day. With no time to spare, they were not even given the promised hot field breakfast.

The two under-strength battalions comprised six companies and four machine gun sections, with nine machine gun teams.

The battalions were commanded by two Harvard and Plattsburgh alumni: Charles Whittlesey, still a gangly, bespectacled lawyer with wire-rim glasses, and George McMurtry, a veteran Rough Rider. Major Whittlesey took command as the senior officer, and Captain McMurtry acted as his executive officer.

Charles was often described as a "stork on stilts," long-legged, thin, and good at organizing. His men saw him as a taskmaster and lovingly nicknamed him "Galloping Charlie" for his gangly stride.

On October 2, 1918, the merged battalions broke through German lines. However, French and American units failed to provide flank security, and Charles and his unit were surrounded in the Charlevaux Ravine, at the base of a steep slope.

The unit quickly organized an oblong defensive position, placing the wounded and command group in the center. Everyone was instructed to dig funk holes. The barrage of mortar, machine gun, and rifle fire from all sides began almost immediately. Over the subsequent five days, gunfire was so intense that soldiers were reluctant to leave, sleeping and relieving themselves in the water-filled holes.

The spot was described as so thickly wooded "you could see no one ten yards away." Germans occupied the high ground, and there was a swamp below them. Germans attacked down the hill, frequently heaving grenades and spraying automatic fire.

Charles's men effectively repelled numerous assaults by the enemy using their 1903 five-shot Springfield and six-shot Enfield rifles and abundant use of Mk 1 fragmentation grenades. Each man carried about 40 clips, or 200 rounds of ammunition. Charles had also effectively deployed their light Chauchat machine guns, provided by the French army. This reliable automatic weapon allowed for a single operator and had twenty-round magazines that could give a firestorm of support.

Germans very effectively employed a short-range 25-cc mortar called the *Minenwerfer*. Its trajectory could destroy funk holes and cripple anyone moving aboveground. Most German soldiers used the highly rated five-shot German Mauser, which some consider the best rifle in the world. The Germans bolstered their rifle fire with MG08 machine guns capable of firing 500 rounds a minute, using the same ammunition as their Mausers.

The Germans also used a highly effective stick-handled grenade called a "potato masher" to concuss American soldiers at close quarters. Combat was so close at one point that Major Whittlesey pulled a grenade's stick handle out of Captain McMurtry's back. By the end of the first day, a third of the unit was killed or wounded.

Under the direst conditions, the unit was pinned down from October 2 to October 7, 1918. There would be no medical treatment for the screaming and dying. Men would be without food for over 100 hours, and a number were killed trying to go to a nearby swamp to fill their canteens. They often resorted to drinking out of mudholes. Whittlesey permitted his soldiers to riffle through dead Germans' packs and clothing for food, but by the fourth day, they mainly ate roots and leaves.

During the Battalion's encirclement, the 50th Aero Squadron conducted the first air re-supply effort in history, flying fifteen missions. Sadly, most supplies missed the drop mark, and the Germans captured them. The two pilots involved were shot down and killed later the same day. Both would be awarded the Medal of Honor.

At one point, the Lost Battalion was bombed by its own artillery, and only Charles's last carrier pigeon named Cher Ami could get through and give accurate coordinates so the shelling would cease. Charles's message said: "We are along the road parallel 276.4. Our artillery is dropping a barrage directly on us. For heaven's sake, stop." (Cher Ami, shot through the breast and leg, still delivered the message, and doctors fashioned a prosthetic leg for the bird, who now rests in the National History Museum in Washington, DC.)

Everyone was sleeping in muddy funk holes with no blankets or overcoats; no fires were allowed. Enemy fire destroyed five of nine machine gun positions as the fighting continued. Before the battle

HEADQUARTERS of 1st Bn 308 Int Form 6-P. S.
TO C.O. 308 Infty
No. 1 Date 3 Oct PIGEON SERVICE Signal Corps, U.S.A. 574
We are being shelled by German artillery
Can we not have artillery support.
Fire is coming from North West
1594
FROM
PLACE 294.6-276.3 TIME 8.50 a. m.
Sender's Signature Whittlesey Major 308
Time of Receipt at Loft No. of Copies sent by P. S. 1

HEADQUARTERS of 2 Form 6-P. S.
TO
No. Date PIGEON SERVICE
open our lines of communication, owing to German Mach. Guns which have filtered them in our rear or come in from the WEST, and we shall not be able to do so without abandoning part of our position, which we hesitate to do, as we understand our mission is to advance.
We have sent 2 pigeon messages today asking that our lines of communication be kept open by units from our rear that we may get food and ammunition, and advance.
86
FROM C.O. 1st Bn 308 Inf
PLACE 94.8-74.5 TIME 4 a.m. p.m.
Sender's Signature Whittlesey Major 308 Inf
Time of Receipt at Loft No. of Copies sent by P.S.

Two of Whittlesey's messages successfully delivered by the Pigeon Service, one of which states, "We are being shelled by German artillery. Can we not have artillery support." *[Courtesy of Special Collections, Williams Libraries, Williams College]*

ceased, the unit was down to two guns.

Charles was everywhere, going on or sending out combat patrols. His men were becoming too weak to bury the dead, so he advised them to throw leaves and brush over them. They salvaged dressings from the dead and reused them on the wounded, or used shirts for bandages. As time passed, with no relief, wounds began to turn gangrenous.

Charles vividly remembered falling asleep next to a young, wounded soldier and waking up with his cheek touching the cold cheek of the dead trooper.

When the situation seemed at its worst, a message to surrender from the German commander was delivered by a prisoner of war under a white flag on October 7. The message read, "We have heard the cries of your wounded. You can't escape. Why do you not surrender, in the name of humanity? Send your reply by a messenger carrying a white flag." Major Whittlesey and McMurtry conferred and rejected any thought of surrender. McMurtry is quoted as saying, "We've got them licked, or they wouldn't have sent this." Charles is alleged to have suggested, "Go to Hell," but in the end they elected to simply not reply, the messenger was allowed to remain, and the Lost Battalion fought on.

One of Charles's last runners did eventually get through, and on October 8, after the Germans had withdrawn, the 307th Regiment appeared to provide relief. The picture that greeted the 307th was bleak: a scarred landscape with trees whittled down from artillery fire, water-filled mortar holes, corpses everywhere, German and American. Broken accoutrements of war were scattered about, discarded rifles, machine guns, and men. They found ripped clothing, riffled uniforms, and knapsacks, including torn boots, some with feet still in them.

The rescuers took care gathering and treating hundreds of wounded soldiers, many now with gangrenous wounds. The men were carried on stretchers and moved to railroad cars, uncovered in the rain, for a bumpy ride to a field hospital.

Fewer than 200 could walk, and most of those ended up in hospitals for some type of treatment. The rest of the group was either killed, wounded, or captured. Charles had lost three-quarters of his men. Casualties numbered over 350 (107 dead, 190 seriously wounded, and 63 missing).

When the battalion had been encircled and cut off, the newspapers had begun weaving the legend of the Lost Battalion. When relieved from duty, Charles learned that over the past five days, the battalion had become a major news story and that he was a national celebrity: The Hero of Charlevaux Woods. His commander, General Alexander, promoted him to Lieutenant Colonel.

When some questioned his loss of troops, General Alexander replied: "Whittlesey's command did what it was told." After the war, General John Pershing, commander of the American Expeditionary Forces, was said to have regarded Whittlesey as one of the three great heroes of the war.

Charles returned to the United States, landing in New York on November 14, 1918, three days after Armistice, and was mustered out of the service on December 5. He was formally awarded the Medal of

Honor in Massachusetts on December 24, 1918, by Major General Clarence R. Edwards on the Boston Common. His other decorations included the French Croix de Guerre. Captain McMurtry was also awarded the Medal of Honor for his actions, separately.

For Charles, the fame he never sought, wanted, or was par-

Both images are conserved at Williams College. *[Courtesy of Special Collections, Williams Libraries, Williams College; and Mike Remillard Photos]*

ticularly suited for was on his doorstep. It was a time of countless interviews. The thirty-four-year-old remained conflicted in his feelings about a battle that saw the loss and suffering of so many of his men. He had advanced, taken the objective, and hung on—just as ordered, but at a huge price.

A survivor's guilt, possibly combined with shell shock, competed with his sense of duty and personal morality. Others criticized his battle efforts as being overzealous or lacking experience.

Upon his return, Charles resumed a private law practice, handling complex cases with his best friend, and maintained a close circle of friends, including his brother Elisha. Charles's family has said that he "never married or engaged to be married," yet he was still sociable, In particular, he enjoyed football games, especially the Army-Navy games.

In October 1918, he was asked to assist in running the Red Cross Roll (second in command), a charitable organization that raised funds to build hospitals for sick and disabled veterans. They wanted to use his well-known name, "Whittlesey and the Lost Battalion," in the efforts to help veterans. Before long, he was asked to be chairman of the organization and continued his practice of visiting hospitals, attending funerals, comforting soldiers, and often found himself listening to someone who had lost a relative in the war.

Veterans continued to seek him out for help in finding a job, a loan, or just to talk. He had a podium in his Red Cross role, and extolled his troops, many of whom had received no official recognition for their bravery and sacrifice. For his men that survived the war, many returned to the same pre-war discrimination that, even after their sacrifice, called into question whether immigrant soldiers were 100% American.

It was a difficult time for veterans. Many had not been paid, were disabled and unable to support families, or had difficulty finding jobs. There was no plan for reintegrating veterans into civilian life, no pensions or benefits, and little medical support for the millions of returning soldiers. They were paid only $1 a day, and that was in arrears. Also, the $60 demobilization bonus was inadequate; getting their old jobs back was not always easy. The government took the stance that veterans were not entitled to more benefits than other workers. Whittlesey's war seemed to have no end. He told a friend, "Not a day goes by, but I hear from some of my old outfit about some sorrow or misfortune. I cannot bear it much more. I want to be left in peace."

In 1921, he was promoted to Colonel, then was asked and

accepted the honor to command the 308th Infantry Regiment. Several months later, he and his comrade George McMurtry were honored guests in the burial of the first unknown soldier at Arlington Cemetery on November 11, 1921.

On November 21, 1921, Charles bought a ticket to Havana on United Fruit liner *Toloa*, departing Saturday the 26th. No one knew he was leaving New York. He told his housekeeper, "I'm going away to be alone for a few days. I am tired."

After boarding that evening, he dined at the captain's table, and after pleasant table conversation and cigars just before midnight, he retired to his room . . . and was never seen again.

When his stateroom was found empty the next morning, there were nine envelopes and a message for Captain Grant indicating Charles had gone over the side of the ship right after his midnight goodbyes. The *Toloa* turned around and spent a half-day searching for Charles to no avail.

Captain Grant contacted the Consul-General in Havana, who dispatched a note to the US Secretary of State and began conducting an investigation. The investigative report stated that Whittlesey "appeared in good health and spirits and entered into a lively discussion on football, manifested an active interest in the Army and Navy football game being played that day . . . . After dinner, he repaired to the smoking salon . . . and talked with (Mr. Wilmot) for more than two hours on a wide range of subjects. He (Mr. Wilmot) noted nothing unusual. . . ."

The report also stated that he "seems certain to have died of his own volition by drowning on the high seas" and noted that the captain was asked to throw all his personal effects over the side, which the captain did not do. The captain did notify Charles's executor, John Bayard Pruyn, a fellow Williams graduate and law partner, who would then begin the notification of family.

When the contents of Charles' letter to Pruyn were learned some years later, it said, "Dear Bayard, Just a note to say goodbye. I'm a misfit by nature and training, and there's an end to it." He apologized for making Pruyn his executor and then mentioned his finances, life insurances, and where his Medal of Honor was stored (in a safe deposit box). At the end of the letter, he didn't explain what he was about to do, just wrote, "I won't try to say anything personal, Bayard, because you and I understand each other."

Then, simply: "Give my love to Edith. As ever, Charles Whittlesey."

It was later determined that he had organized his affairs, left his office work in good order, and paid his rent in advance. Charles had left letters for his parents, brothers (Elisha and Melzar), an uncle, and close friends, including George McMurtry. They all refused to make the letters public. Poignantly, Charles left to McMurtry the document of the Germans' request for surrender.

Just three years after winning the Medal of Honor, Charles had jumped into the ocean late at night somewhere between the United States and Cuba. His death was described as "drowning at sea by own intent." His body was never found.

In tribute to Charles, on December 11, 1921, three thousand people crowded the State Armory in his hometown of Pittsfield, Massachusetts, to mourn the loss of their hero. An empty flag-draped caisson was the center of attention. Charles was again honored when a dedication tablet was placed in the Massachusetts State House in his parents' presence in December 1924.

On a different note, that same year Charles's mother, in remembrance of her son, obtained a license plate from the Bureau of Motor Vehicles with numbers corresponding to her son's division and regiment 77308 (308th Infantry, 77th Regiment).

Charles had sympathized with his men, who struggled with mental and physical health issues and unemployment. Almost certainly, he suffered from depression and untreated shell shock (today known as PTSD). Although he always seemed to keep his nerves under control, he struggled with his feelings against war, its devastation on his men, and his patriotic duty.

Losing so many men in a battle that had a dubious outcome haunted him, as did his nightmares, often remembering sleeping cheek-to-cheek with one of his dying soldiers. After much analysis, it is believed he could not shake its effects. Years after the firing had ceased, Charles too had become a casualty of war.

Private First Class John C. Hitchcock
Silver Star
WORLD WAR II

They were ahead of Company C, tasked as reconnaissance scouts, searching and probing for German ambushes. John Hitchcock and his fellow scouts had experienced countless close calls for almost a month in continuous combat. As part of the Rainbow Division, their company was storming through Germany.

Even in the last days of the war, the "Rainbowmen" were facing a fanatical enemy in the diehard Waffen-SS and Hitler Youth, who were ordered to (and determined to) resist to the last man. The scouts knew this as they rapidly approached the outskirts of the tiny village of Döckingen, slightly ahead of Company C, armed with M-1 rifles, grenades, and .45-caliber pistols. Their thoughts were suddenly interrupted by an intense artillery barrage. A hailstorm of machine gun and small arms fire enveloped them. It was April 1945, the long war was coming to an end, and no one wanted to die now.

Early Life

JOHN'S STORY BEGINS IN Pittsfield, Massachusetts, with his birth on December 6, 1925, to Wilfred M. and Mary Cleveland (Leslie) Hitchcock. Shortly after his birth, the family moved to nearby Windsor, where John grew up. In Windsor, John learned to ski at the age of seven, igniting an avocation that stayed with him for the rest of his life.

Wilfred, his dad, worked in stationery production at the Crane

Company and, outside work, tended a small farm that mainly produced turnips and potatoes. He and Mary had five children: three boys and two girls, one every two years. Mary was regarded as quite an independent lady. For example, at an early age, her father had taught her how to use a plow. Even more revolutionary, before her marriage, she had attended art school in Boston. Mary was active in her community (described by some as "feisty") and, when in her nineties, was voted Berkshire County Woman of the Year.

At one point, with seven people in the household, it was decided that John, now ten years old, would live with his maternal grandparents, J. Frank and Libbie Leslie, who lived nearby. Not long after John moved in, his grandmother died, and the move took on a more permanent nature so John could help his grandfather with the farm work and, in some ways, be a companion for him. There was plenty of work. Frank ran a commercial Christmas tree farm, was involved in logging, and sold lightning rods. He also made time for local civic activities.

John enjoyed living with Grandpa Leslie. He attended the local Crane Community School and was always busy directly after school with farm work, 4-H projects, and homework. The school was located halfway between his parents' and grandparents' houses, so he got to see his family frequently. During his early school years, John's left-handedness was noted, and as was the practice at the time, his teacher spent considerable time correcting this "deficiency" until he wrote properly with his right hand.

John attended Dalton High School and was cast in a few plays between working on his father's and grandfather's farms and delivering potatoes for them to different markets. His nighttime entertainment consisted of listening to his grandfather's storytelling, a practice he carried on, day and night, in his adult life.

John graduated from high school in June 1943 amidst World War II. He knew months in advance that he wanted to be a pilot just like his older brother, so the day after graduation, he headed to Boston to test for the Navy's flight school. Concerned about his marginal eyesight, John had been on a carrot-eating binge. Even after weeks of ingesting carrots, he failed the exam, and his skin acquired a temporary orange tint. He never relished eating carrots again, but shortly afterward, John entered the US Army at age seventeen.

Joining the Army Reserve, John was selected to attend the Army's Specialized Training Program (ASTP) at the University of Maine in 1943. He would complete the six-month program in March 1944, embracing the physical challenges of performing hundreds of sit-ups, swimming across a wide river, and dealing with older and bigger recruits in combat wrestling.

Of note, ASTP was a World War II training program to meet the personnel demands for soldiers with technical skills. A network of colleges and universities offered technical coursework in such areas as engineering and medicine. Participants needed to be on active duty, pass a qualifying test, and be ranked a private. The almost sixty-hour workweek included twenty-four hours of classroom activity, twenty-four of study, six of physical education, and five of military instruction. Eventually, the ASTP program was greatly diminished, due to the need for combat infantrymen in Europe.

John's next assignment was three months of basic army combat training at Fort Belvoir, Virginia, and then ten more weeks of electrical trades training in New York City. The combat technical training taught John how to build bridges and roads, clear minefields, and demolish pillboxes and wire entanglements. When John was training in New York, he commuted home on weekends. On his first weekend home, a local minister visited the family to break the news that his older brother Jim Hitchcock had been shot down over Italy and was being held in a German prisoner of war camp near Munich, Germany.

John's training continued into September 1944 when he began ten weeks of Advanced Infantry Training at Camp Gordon, Georgia. At Camp Gordon, the Army's largest training facility, tens of thousands of soldiers were trained in weapons handling, tactics, and drills.

Successfully completing infantry training, John embarked with a convoy of ships leaving New York City in late February 1945. He had been tasked with helping others board and settle in and was

John as a very young soldier. *[Courtesy of Hitchcock family]*

amazed at the streams of young men entering the vessels headed for Europe. The trip took almost a month, and some boredom was broken by reading, endless card games, and the persistent and delightful rumor among the troops that one of the ships in their convoy was carrying WACs (Women Army Corps).

John landed at Le Havre, France, in late March 1945 and was immediately assigned to the 42nd Rainbow Infantry Division as a replacement. The division had landed in Marseille in December 1944 and been fighting its way across Europe. Replacements were needed, badly.

The 42nd Rainbow Division, whose motto is "Never Forget," was created in World War I and sent to fight in France. Its nickname was derived from its broad range of National Guard units from 26 states and the District of Columbia. It represented a cross section of the United States, and its soldiers became known as "Rainbowmen."

The division was reactivated in World War II, landing in Marseille, and ultimately traveled over 1600 kilometers through France, Germany, and Austria. The division was composed of 14,000 men in three infantry regiments (222nd, 232nd, and 242nd).

After landing and being assigned to the 42nd Division, a quick orientation was required for this boy from the town of Windsor, Massachusetts, with its population of little more than 300 souls. In discussions with other replacements, all seeking optimum choices in a risky environment, somehow John became assigned as a scout, a perilous job in the best of circumstances.

He was assigned to Company C of the 1st Battalion of the 232nd Regiment. There were supposed to be 150 men in his company, but casualties severely impacted this number. He joined the company in late March, and the ensuing fifty days were the division's most intensive period of combat. John's division engaged in a running battle, moving nonstop from the center of Germany southeast toward the Austrian border.

Many of the Rainbowmen were replacements, mostly teenagers. Eighteen- and nineteen-year-old boys were thrown into this deadly, front-line, last-gasp maelstrom, something that would haunt them the rest of their lives. John himself was nineteen. He wrote to his family once to warn them that due to moving around a lot in Germany, they may not hear from him as much. He also mentioned being upset that while getting ready for battle he had to throw out his dictionary and chess board. He asked his mom for a book that contained some of Shakespeare's writings.

On March 30, 1945, the Rainbow Division crossed the Rhine River and started passing through towns destroyed by the Army Air Force. It was difficult to absorb the devastation seen by the young Rainbowmen: bloated, eviscerated bodies along the road, screaming, wounded livestock, shattered vehicles, orphaned children, desperate, hungry civilians, German soldiers hanging from lampposts (alleged deserters). It was Orwellian.

As a scout, John led Company C through near-daily action against the Germans, fighting in little-known, bombed-out towns, villages, and cities with names like Würzburg, Schweinfurt, Fürth, Döckingen, and Nürnberg, eventually leading to Munich.

John and the scouts would be just far enough ahead of the

company to ensure it avoided enemy ambushes. With the scouts in the lead, Company C advanced on foot or by truck from one occupied town to the next, freeing them from enemy forces. Small towns and villages were often intact, but large cities had been devastated by bombing. The scouts were never sure what their reception would be, but often it was a fusillade of bullets from determined fanatics dug into collapsed buildings.

The battalions quickly fought for and secured one town after another, moving on to the next one. Some of the names they never knew; sometimes, the soldiers weren't sure where they were. They just moved on with little rest, no hot meals, and some cold K-rations.

The Germans had prepared a town-by-town defense for the Rainbowmen. Small towns between their major objectives were often defended, making progress excruciatingly slow. Roads were mined, and diehard German soldiers were ordered to fight to their deaths. Hitler Youth between the ages of thirteen and fifteen, armed with Panzerfaust rockets, attacked Allied troops whenever an opportunity presented.

In his unpublished memoirs, John remembers many harrowing close calls during the division's final charge across Germany.

He and his fellow scouts leading from the front were the frequent targets of snipers, artillery, and missed land mines. In reflections shared with his family, he remembers one night staying "on the second floor of a barn . . . near an open window, and suddenly felt bullets pass by [my] face . . . and saw tracings of their paths in the dusty air.

"Another time, in a town, [I was] asleep on a chair, with [my] feet drying in an open stove door, when a sniper fired through the kitchen window . . . another bullet whizzing by [my] face."

During a day operation, German Tiger tanks fast approached John, "and others had to lie very flat in the ruts of a muddy road . . . with the plan being someone in the unit would disable the tanks as they got closer . . . luckily someone did."

Once, he and two scouts were surprised by a large group of Germans coming down a trail on a murderous rampage, killing everything in sight. The scouts hid in a slight depression and covered themselves with leaves, barely avoiding being discovered.

The first major operation for John began on April 2, 1945, at the city of Würzburg, most of which had been destroyed by bombing. The Germans were dug deeply into the rubble. After intense fighting, Company C entered the city on April 4, clearing sections of the vast stone wasteland. It was exhausting, climbing over half walls, shooting through windows, throwing grenades through doorways, and waiting for the inevitable sniper shot from a Mauser 98, the Germans' most accurate sniper rifle. Yet they secured the city.

Casualties were so high that some infantrymen started to think the only way out of their rifle company was to be dead or wounded.

There was little rest before Company C moved on to their next objective: Schweinfurt, thirty miles away, another decimated city with more house-to-house fighting. This deadly routine would repeat itself in towns and hamlets throughout April.

The division halted on April 13, 1945. The men learned that President Roosevelt had died, and they held a spontaneous memorial service. Photographs of the time show soldiers from the 242nd Regiment gathered for prayers before the flags of all the states amidst the ruins of Schweinfurt.

Fighting quickly resumed, and the Rainbowmen crossed the Danube River. The three regiments continued to seize the cities and inch toward their goal of the Austrian border, still hundreds of miles away.

April 23 was an especially difficult day for Company C, 1st Battalion. With scouts in the lead, Company C had the goal of capturing the tiny village of Döckingen, just adjacent to the medieval town of Rothenburg. The first battalion approached the town from high ground, and Company C engaged SS troops in a running skirmish. They could see a little girl lying in the street, shot by the SS when she announced the Americans were coming. German artillery and small arms fire rained down on the infantrymen as they approached.

The platoon commander sent out a scout patrol to the German flank to act as spotters to help knock out German artillery. The US batteries succeeded, after 15 minutes of counterfire. Then the company moved into town and eliminated the remaining machine gun nests and small arms fire.

It was during one of these actions that John, while acting as a lead scout for his platoon, came under fire by a German machine gun and small arms. He was wounded in the right elbow. Despite the wound, he advanced on a machine gun position and destroyed it with grenades. Before seeking medical aid, he continued "to engage the Germans for several more hours in one of the last major engagements of the war."

Later, he and other wounded were gathered at a farmhouse, treated by medics, and John spent several days in the hospital before returning to duty.

In December 1945, John Hitchcock was awarded the Silver Star by General Harry Collins, Rainbow Division commander, for "Gallantry in action on April 23, 1945, near Döckingen, Germany." The citation said that Private Hitchcock's extreme courage under fire was instrumental in ensuring the rapid advance of his company.

In congratulatory words to John's regiment, the Commanding General noted the following on April 23: "We advanced 13 miles thru mud and snow, much of the way under enemy artillery fire, captured eight towns, one artillery piece, and 50 prisoners, knocked out another artillery piece and two troop wagons, blew up a gas dump and at dusk captured intact a bridge that the Division needed for a crossing and secured a string bridgehead that same night. To accomplish this, we advanced 3 miles beyond the forward elements of the Division and were surrounded by enemy troops." They lost a lot of men that day.

Back on the line, it was only days later, on April 29, when other companies in John's regiment liberated the oldest Nazi concentration camp, the infamous Dachau. John's fellow soldiers found 32,000 walking skeletal prisoners, some barely breathing, others actively

HEADQUARTERS 42ND (RAINBOW) INFANTRY DIVISION
OFFICE OF THE COMMANDING GENERAL

[illegible] December 1945

CITATION

AWARD OF THE SILVER STAR

By direction of the President, under the provisions of Army Regulations 600-45, [illegible] September 1943, as amended, the Silver Star is awarded to:

JOHN [illegible] HITCHCOCK

[illegible], Private, Infantry, Company [illegible], 232nd Infantry for gallantry in action on 29 April 1945 near [illegible], Germany.

When his platoon was surprised by enemy machine gun and small arms fire, Private Hitchcock was wounded while acting as a lead scout. Despite his wound, he advanced on one machine gun position and destroyed it with grenades. Before seeking medical aid, he continued to attack until his company had secured their positions. Private Hitchcock's extreme courage under fire was instrumental in ensuring the rapid advance of his company. Entered military service from Springfield, Massachusetts.

[illegible]
Major General, USA
Commanding

An image of the Silver Star commendation, along with awards and citations, framed by his family. *[Courtesy of Hitchcock family]*

dying, and thousands of disintegrating bodies in boxcars and incinerators. Death permeated the air. It was a sight the soldiers never forgot, and the word quickly spread throughout the regiment about what they had found.

The Rainbowmen continued, entering Munich on April 30 with little resistance. Entire German groups surrendered and lined the road; for them, the war was over. The Rainbowmen continued to the Austrian border and crossed over on May 4. The Germans officially surrendered on May 7, 1945.

John's older brother Jim, the fighter pilot who had previously been shot down in Italy, was in a prisoner of war camp in Munich when John passed through the town, unaware of his proximity. Fortunately, Jim was safely released at the end of the war.

The soldiers of the 42nd Division had an impressive 106-day combat record in 1945, and the men were exhausted. They had captured the key German cities of Würzburg, Schweinfurt, Fürth, Nürnberg, and Munich (and scores of towns and villages in between). The 42nd was the first unit in its corps to cross the Danube River, the first into the Dachau death camp, and the first into Hitler's retreat known as Eagle's Nest. During this period, they were credited with capturing 59,000 prisoners, while sustaining thousands of their own casualties.

The division received much-needed rest when it set up camp near Salzburg, close to the famed ski resort of Kitzbühel, and John was tasked with building a sniper range in preparation for the invasion of Japan. The division also processed thousands of German soldiers, maintained roadblocks, collected weapons, and screened and searched for SS who may have committed war crimes.

In one of John's recollections, he mentions being put in charge of improving the poor food offered to the soldiers at the mess hall. John took charge, commandeered a small Inn, and chose an experienced staff of chefs and head waiters from a select group of prisoners. By trading army supplies for fresh produce and using a bread truck to gather food from local citizens and farmers, it did not take long for the Inn to be the place to eat.

John's greatest enjoyment after the past months of harrowing combat came on the weekends, when he had time to enjoy some of the world's best skiing at Kitzbühel, Gaisberg, Zell am Zee, and St. Anton. In a letter dated October 1945, he asks for a photo booklet of the Berkshires to show Salzburg residents there is more to the world than just the Alps! Though skiing is his greatest pastime, at one point he also took up painting with watercolors and charcoal.

Finally, John returned to the United States in June 1946. He was honorably discharged at Fort Dix, New Jersey. Three years of active duty had produced a decorated hero for the village of Windsor,

Massachusetts. John had earned a Silver Star, a Purple Heart, the Combat Infantrymen's Badge, the World War II Victory Medal, a Good Conduct Medal, the Army of Occupation Medal, and the European, African, and Middle Eastern Medal.

In John's absence, in order to save him a place, his brother enrolled him at Williams College, and he began in 1946, majoring in English literature. John supported himself by working on a tree-trimming crew and teaching skiing at nearby resorts. He returned briefly to Europe to rekindle friendships and enjoy the great skiing venues.

Photo taken upon returning to Switzerland in 1948 for a postwar visit. *[Courtesy of Hitchcock family]*

Some years prior, John's brother had married a young woman named Ruth Milne, and when John, while in uniform, met his brother's in-laws, he caught the eye of Ruth's lovely sister Ann. John and Ann were married in 1948, and in 1950, John graduated from Williams with a Bachelor of Arts degree.

In the early years of their marriage, John and Ann enjoyed traveling cross country. He worked first in Idaho for the *Denver Post*, then had a leadership role with the Berthoud Pass Ski Patrol, until he broke his leg while skiing at Aspen. Next, he and Ann returned to the Berkshires, where John worked at a small radio station in North Adams before they headed off to New Hampshire and a staff reporting job at the *Portsmouth Herald*.

Returning to the area, John and Ann settled in Williamstown with their three daughters, Sally, Amy, and Jan. He worked for

The Berkshire Eagle, where one of his responsibilities, as ski editor, permitted him to cover skiing for Massachusetts, Vermont, the Adirondacks, and Connecticut. John also worked for several years as Williams College Assistant News Director.

For almost thirty years, John used his prolific journalistic skills as a Bureau Chief for the *Springfield Union* newspaper. Not a person traditionally driven by money, John once refused a senior editorship position with a ski magazine that would have involved a move to New York City; he decided to stay in his beloved Berkshires.

Crafting his stories in the long afternoons and evenings for the next morning's release allowed him the freedom to pursue outdoor activities each morning. John loved to hike, golf, and play tennis, but skiing was his passion. It seems to have grown from his early days in Windsor and accelerated during his time in Salzburg.

He had a knack for organizing and taking charge of programs, and in the 1960s, he founded the All-American Ski Camp in Montana, with Spaulding Company as a sponsor. The Montana camp was a summer training program for aspiring young racers; he also offered an autumn training opportunity in Switzerland.

John played many different roles at Berkshire ski areas teaching downhill and cross-country skiing at Dutch Hill, Bousquet, Brodie Mountain, Jiminy Peak, and Petersburg Pass. He operated cross-country ski centers, was active in other local sporting events, and consulted on public relations for tennis and golf facilities, including Waubeeka golf links.

At different times, he was ski school director at Jiminy Peak and Petersburg Pass, sharing his love of skiing with hundreds of others. Outside of his official duties, though, he could be a daredevil, like skiing on one ski when he had broken a leg (from skiing), trying out roller skating down a very steep hill on a long street, or testing his brakes on icy parking lots at ski areas.

With his journalistic skills and skiing experiences, John was considered by many to be a ski writing pioneer, which allowed him to write for his employer and numerous magazines, including

John Hitchcock in the 1980s on Brodie Mountain, showing his clear zest for life. *[Courtesy of Hitchcock family]*

Sports Illustrated, *Ski*, *Skiing*, and *Ski America*. John also co-authored *Making it: A Guide to Cross Country Skiing, Snowshoeing, and Survival.*

He also wrote for numerous ski newsletters, articles for the Associated Press, and often a series of articles for ski periodicals. John acted as a ski consultant a number of times. He covered the sport throughout the world, including the Squaw Valley Olympics. John was an accomplished photographer whose ski photos were used by the Associated Press worldwide.

He loved to travel, especially if there were ski areas involved, and did so regularly through his activities as a sportswriter with his wife and daughters. When on trips together or at home, he was always encouraging his daughters' engagement with sports. Travel for skiing ranged throughout the United States, Canada, South America, and Europe.

His daughters remember an energetic dad who always encouraged outdoor activities. Mornings might be filled with the smell of strong coffee, melting ski wax, or their dad cooking from his cast-iron skillet, making omelets with veggies, black olives, and cheese. (Growing up with his grandfather, it was his only cooking implement.) There was no sleeping in . . . too much to do! Besides, there was always "twig patrol" when the yard needed picking up. Never a dull moment with Dad.

John and his daughters spent a lot of time hiking. Later in life, they remembered how very vigilant he had been when they were in the woods and attributed it to his time as a scout in Europe. They did not know that he had experienced nightmares when first returning home from the war, and that fireworks and guns made him very uncomfortable, signs we understand today as PTSD.

Family members often thought that when he was skiing, it was a happy time and place for him, and some surmise it acted as therapy. Late in his life, asked by a rehab facility's activity director if he had any religious preferences, he responded, "Skiing and tennis."

After the war, John, through his prolific and engaging journalism, had a tremendous influence on promoting and coordinating outdoor life. When skiing is discussed in the Berkshires or southern Vermont, it is often in connection with John Hitchcock. One of his family members recalls John having t-shirts that read HITCHCOCK'S THE NAME—SKIING'S THE GAME.

Although John stopped skiing in his late seventies, he continued to write weekly articles for the *Springfield Republican*. Subsequently, he wrote the "Snow Sports" and a recreation column for the *Advocate* well into his eighties.

That young boy who grew up alongside his grandfather so many years ago in the small village of Windsor became a war hero, a journalist, and an outdoorsman, all natural outpourings of his energy, zest, and love of the outdoors. John Hitchcock's optimism, gregariousness, and ability to connect with people created a lasting legacy that lives on, not least in his beloved daughters, but also in those many people he touched throughout the Berkshires and beyond.

1st Lieutenant Marc H. Jaffe

Battles of Peleliu & Okinawa / Bronze Star

WORLD WAR II

This Harvard literature graduate (Class of '42) who wanted to be a combat correspondent instead found himself on the front lines, leading men through two of the most vicious battles of World War II: Peleliu and Okinawa.

Marc Jaffe survived three years of active duty as a Marine, then established himself in the world of publishing. He was described by The New York Times *as "one of the most highly regarded editors in the mass-market paperback field" and a "shrewd judge of mass taste." Marc continued to be highly successful during what has been called "the golden age of publishing" and spent years atop the New York literary establishment. In July of 1961, he joined Bantam Books, which became one of the most formidable mass-market paperback publishers of its day, and by 1978 he was named its President. He has now entered his second century, enjoying life in Williamstown, Massachusetts, and remains active as a freelance editor and publishing consultant.*

Early Life

MARCUS HENRY JAFFE, FAMILIARLY known as "Marc," was born in Philadelphia, Pennsylvania, on November 6, 1921. His father, Dr. Samuel Jaffe, had been a major in the Army during World War I and served in France. A general practitioner, Dr. Jaffe was Chief of the Eastern Division of the Albert Einstein Medical Center and also taught at Jefferson Medical Hospital.

Marc's mother, Lily Bailey Jaffe, was brought up on a farm in New Jersey. She was a schoolteacher and, early in her career, managed a school for "wayward" girls. Lily became a social worker and was active in the Auxiliary of the Einstein Medical Center. Lily and Samuel married in 1916 and moved into their house on Pine Street soon after.

Samuel had emigrated from Lithuania via Ellis Island, and Lily's father from Odesa, Ukraine. The young couple had three boys: David, Marcus, and Robert. All three would serve in World War II. Both of Marc's brothers became doctors.

The boys were raised in what is now called Philadelphia's Society Hill neighborhood, attending Oak Lane Country Day from elementary through high school. The school was owned by Temple University as a training ground for teachers; Marc said it provided "a great education" through its progressive approach.

The family was busy. With two working parents, they lived a good middle-class life on their quiet block across the street from the historic St. Peter's Church. Marc remembers his mother instructing the boys that they had to eat everything on their plates, even her milk soup, a pallid concoction of milk, carrots, and celery.

MARC HENRY JAFFE

Born November 6, 1921, in Philadelphia, Pennsylvania. Prepared at Oak Lane Country Day. Home address: 309 Pine Street, Philadelphia, Pennsylvania. Adams House. House football (2); soccer squad (1). *Red Book* (1). Phillips Brooks House (1). Student Union (3, 4). *Progressive* (3, 4). House dramatics (2, 3). John Harvard Scholarship. Field of concentration: History and Literature.

Harvard yearbook photo. *[Courtesy: HUD 342.04 Harvard University Archives]*

Marc, a bright student, was accepted at Harvard at the age of sixteen and entered as a freshman in 1938. He decided to major in American history and literature and, though he devoted a major portion of his time to study, did play soccer his freshman year. Marc was also a yearbook contributor, participated in the Phillips Brooks House community efforts, and was involved in his house's dramatic shows. His plans for the future were dramatically altered when the Japanese attacked Pearl Harbor on December 7, 1941, in the middle of his senior year.

With the war going on and vivid accounts coming in from the front, Marc decided to become a war correspondent like John Hersey and James Aldridge from *The New York Times* or Ernie Pyle from Scripps-Howard newspapers. Marc traveled to Washington, DC, and talked with the Marine Corps public relations office. The officer he met with told him the Marines had plenty of combat correspondents, but there was a need for field officers. Not wishing to chance a future as an Army draftee, he volunteered to enlist in the Marines as an officer candidate.

Military Service

Marc went on active duty in the Marine Corps in February 1943. He began, as many Marines do, in recruit training at Parris Island, South Carolina. He remembers that as his unit was arriving at Parris Island, a truckload of Marines yelled out, "You'll be sorry!" After six weeks of boot camp, he moved north to the Officer Candidate School at Quantico, Virginia, for a final ten weeks of training.

Upon graduation, Marc was quickly transferred and assigned to the 33rd Replacement Battalion at Camp Lejeune. The unit was briefly involved in field exercises, then traveled via train to San Diego for embarkation overseas. Assigned to the 1st Marine Division, Marc departed the west coast on the USS *Orizaba*, a former cruise ship, now a troop transport. A week or two later it dropped anchor at New Caledonia, where a stopover of a few weeks gave the replacements at least a chance to admire the beautiful island.

Finally assigned to the 2nd Battalion, 1st Marine Regiment, Marc continued his travels to Cape Gloucester, New Britain, where he trained as a forward observer for the 81-mm mortar platoon, part of the battalion's weapons company. Up to this point, he had received little forward observer training. He also participated in a few jungle patrols, although most of Cape Gloucester had been pacified.

Lieutenant Marc Jaffe in the Pacific Theater with an 81-mm mortar. *[Courtesy of Marc Jaffe]*

The battalion then sailed to Pavuvu, one of the largest of the Russell Islands, with an overnight stop in Brisbane, Australia, where Marc remembers enjoying a milkshake at a dairy bar near the wharf. Pavuvu, the temporary headquarters for the 1st Marine Division, was an island, much of which was planted with coconut palms, located sixty miles from Guadalcanal. Pavuvu served as a staging area for the division's next battle: Peleliu.

Peleliu

Peleliu Island is six miles long and two miles wide. The 1st Marine Division, including Marc's regiment, the 1st Marines, had been given the responsibility of overwhelming the occupying Japanese forces. The three Marine regiments would be up against 10,000 entrenched Japanese defenders.

The regiments planned to land on Peleliu's southern beaches, with an immediate goal of capturing the nearby airfield, then surge north to capture the remainder of the island.

Peleliu was subject to an intense three-day naval bombardment, and the 1st Marine Division's Commanding General, agreeing with Naval Command in Hawaii, predicted it would be secured in four days. In the end, it took over two months to complete the mission. Unbeknownst to the Americans, the Japanese had changed their defense tactics, eliminating wasteful "banzai" attacks (all-out suicidal attacks accomplishing little militarily) and instead fortified the coral ridges not far from the beach area that would provide a stubborn defense as the Marines proceeded inland. At one point, the US Naval Commander reported that they had run out of targets, not realizing that most of the Japanese underground fortifications had escaped unscathed.

The Japanese commander had constructed a heavily fortified, interconnected bunker, cave, and underground network, often with sliding armored doors on tracks and multiple openings to protect their artillery and mortars. The island's central mountain elevation provided devastating fire superiority in support of the island's defense.

Several weeks ahead of the planned beach assault, the Marines had boarded LSTs (landing ship tanks) and departed from Pavuvu for the long trip to Peleliu. Each LST carried numerous amtracs (amphibious vehicles) to ferry the Marines to shore for the assault.

At 8:30 a.m. on September 15, 1944, after a steak and egg breakfast, Marc, his radio man, and the platoon they were with descended from the LST's main deck to their amtracs. The LST's vertical doors opened, and the amtracs splashed into the water. With this,

Operation Stalemate II began. Marc was in the first wave along with hundreds of other Marines.

Japanese fire began almost immediately, as artillery and mortars tried to destroy the invading amtracs. The three regiments, 1st, 5th, and 7th, landed side by side over a 2200-yard area in support of each other. The landing zones were designated White 1 and 2, and Orange 1, 2, and 3. Marc's regiment landed on the northernmost beaches, White 1 and 2.

Even now, Marc vividly remembers his first few moments on the beach: he was crawling, wet, face-down, with hot sand clinging to him, bullets flying everywhere, Marines falling and crying out for help. For just a few seconds, it seemed to him that he was watching a movie. A 30-foot coral promontory flanked the beach, providing devastating enfilade machine gun and cannon fire, crippling the landing parties.

This promontory became known as "the Point" by the Marines, and the Japanese 47-mm guns and 20-mm cannons perched there destroyed dozens of landing vehicles. The guns were also supported by mortars and both heavy and light machine guns.

The Marines faced Japanese machine guns, universally called Nambus after their designer Kijiro Nambu. The guns were rugged and reliable automatic weapons using a 7.7-mm cartridge, capable of spraying hundreds of rounds a minute for hundreds of yards, covering most of the landing area. Company K, which finally secured "the Point," suffered 157 casualties out of approximately 200 Marines. Close quarter combat eventually destroyed the Japanese pillboxes and heavily fortified bunkers.

Marc later told *The Berkshire Eagle*, "We had at least 26 out of our platoon of forty Marines that I went in with either wounded or killed in the first half-hour or so on the beach." There were also many heat exhaustion casualties from the 100-plus degree temperatures and lack of potable drinking water.

Marc's job as a Forward Observer (FO) was dangerous. Often under fire, he needed to be close to the enemy to see and pick out targets for the mortar teams and yet cautious enough with his directions to not call fire down on friendly units (including himself). His wireman would run phone wire from the observer back to the 81-mm mortar crew, and they used field phones, not radios, to communicate. Battery-powered and hand-crank field phones were more reliable in humid jungle conditions.

When Marc landed on the beach, he and his wireman, surrounded by the surviving riflemen, moved inland quickly and, once the 81-mm mortars were in place, began calling back fire missions.

As they approached their first objective, the airfield, they could see in the distance a number of Japanese tanks scurrying about. Then suddenly the tanks came straight at them. After taking cover in a shell hole, they heard the tanks pass within feet of their position. It appeared the tanks were focusing on the waves of Marines on the beach. The tanks were, in all likelihood, Japanese 95 light tanks with three-man crews, sporting a 37-mm cannon and two machine guns. Fortunately, they were quickly disabled and blown up by Marines with bazookas. It was a close call.

Armed with a .45-caliber pistol and an M-1 carbine, Marc and his wireman crawled forward through destroyed vegetation, shredded woods, and a swampy area, always headed toward the objective: the airfield. Day one found them at the airfield's edge. The Marines captured the airfield on day two and began a treacherous climb up rocky coral ridges.

The enemy used an array of heavy weapons in the ridges, including a large 320-mm mortar. Its screaming, 600-pound projectile was fearsome, killing Marines and making craters 8 feet deep by 15 feet wide. The Japanese also used a 150-mm mortar with a range of over two miles. All during the early days of the battle, the Marines were under artillery and mortar fire. The mortars were difficult to locate, embedded as they were in caves and with their concealing steel doors.

On one ridge, the captain of Company E, Joe Gayle, who was right next to Marc, was hit in the neck with shrapnel. Marc reacted quickly, putting pressure on the deep wound to stop the bleeding until a corpsman arrived. Many years later, at the battalion's only reunion, Captain Gayle introduced Marc to his wife as the man who saved his life. As Marc told *The Berkshire Eagle*, "I think saving a man's life was the most important thing I did my whole military career" (10-26-21).

After fierce fighting and constant bombardment for days, Marc was brought to the battalion aid station and transferred to the nearby hospital ship, suffering from battle fatigue. He returned to battle the next day. The *Philadelphia Inquirer* reported (inaccurately) that Marc "was wounded October 1 on Peleliu Island but has returned to action with the Marines . . . . [He] was on [a] 36-hour special mission" (11-21-1944).

In the first seven days, the 1st Marine Division had seized all the terrain of strategic value. Prolonged combat, heat, exhaustion, Japanese raids, and rugged terrain had depleted Marc's 1st Marine Regiment. It suffered 1749 casualties, considered high given that each regiment started with approximately 3000 men. (Overall, the casualty rate for the Marines and Army on Peleliu, a total of 28,000 troops, was 35% with 1800 killed and 8000 wounded.)

Operation Stalemate II was controversial in the States. The island was seen as having little value with a very high casualty rate, exceeding all other amphibious operations. General MacArthur wanted to retake the Philippines and said the capture of Peleliu would protect his left flank. Navy Commanders favored bypassing the Philippines. MacArthur won the argument. Many would argue the Battle of Midway three months earlier had decimated the Japanese Navy, and it posed little threat to MacArthur. Others have said the operation was already scheduled, so it was easier to go ahead as the casualties were expected to be light.

Peleliu was officially secured on November 27, 1944. (Of note, one Japanese lieutenant with approximately 30 infantry soldiers and sailors held out in caves until April 1947, when convinced to surrender.)

Peleliu did give the Marines valuable experience in assaulting fortified positions that would serve them well in their next battle—Okinawa. Eight Congressional Medals of Honor were awarded to Marines who fought on Peleliu, five posthumously. The 1st Division also received a Presidential Unit Citation.

Marc's 2nd Battalion 1st Marines returned to Pavuvu for rest and refitting. When the transport dropped anchor, the desolate island never looked so good. The division spent months recovering from the Peleliu battle and training a large number of replacements. It also conducted demanding field exercises preparing the troops for Okinawa.

The battle for Okinawa would be the largest amphibious assault in the Pacific Theater of World War II. In contrast to the original expectation of the Peleliu episode, it was anticipated that this campaign would be costly, as Allied forces would be landing on Japanese territory and closing in on Japan itself, only 350 miles away.

The 1st Marine Division left Pavuvu in mid-March for Operation Iceberg, the assault on Okinawa, which would begin on April 1, 1945.

Okinawa

By now a 1st Lieutenant, Marc found himself in the third wave as they climbed down cargo nets to the Higgins boats that would take them to the assembly area, where they were transferred to amtracs for the landing. No longer a forward observer, he had been promoted to Executive Officer of Company F and waded ashore with it. (Not long after, Marc would become Company G commander.)

The Okinawa landing was the opposite of Peleliu, in that other than sniper and occasional mortar fire, the assault was unopposed. Orders were to move inland in order to cut the island in two.

Okinawa is much larger than Peleliu, sixty miles long and up to eighteen miles wide. At that time, it was defended by over 100,000 Japanese troops and conscripts. Marc describes the action on

Okinawa as obviously very different from that on Peleliu. Japanese "gave up" the beaches, so the first few weeks were easier. Even so, after days of rain and slogging through mud, the newly appointed Commander of Company G dug a slit trench and pulled his poncho over his head in the pouring rain. He remembers saying to himself, "If I get through this, I can get through anything."

The 1st Division Marines, after spending most of April mopping up the island's middle, then headed south. Conditions were insufferable, with May and June miserably chilly and rainy. The monsoon rains began making roads muddy and impassable. Sleeping meant huddling in foxholes filled with water, with a poncho as a roof. As Marc's unit moved south, fighting continued to worsen, as did conditions. There were difficult tactical and medical situations, slippery slopes, constant rain, and unburied enemy bodies.

The 1500-yard Kunishi Ridge dominated the southern end of the island, with Japanese embedded in honeycombed caves on both sides. Marc's unit fought daily skirmishes on various hills and ridges, displacing dug-in Japanese. He remembers explicitly his company attacking Hill 69, just before the formidable Kunishi escarpment, with bullets coming down on all sides.

The prelude to Kunishi, including the taking of Hill 69, involved vicious fighting with high casualties. The ridge itself caused difficulty for the Army and Navy's big guns, and the inaccurate fire resulted in many "friendly fire" casualties. But the Marines held on. Finally, Lt. Jaffe's decimated company was relieved in the middle of a mid-June night. Company G and the rest of the battalion was brought to the beautiful northern part of the island where hostilities had ceased. It was peaceful and quiet. There, Company G rested and refitted, preparing for the invasion of Japan.

Marc earned the Bronze Star for his leadership from June 1 to June 15, 1945, which included a successful nighttime attack on a ridge and holding his position when his company had been cut off from all other units (*Berkshire Eagle* 10-26-21).

The 81-day battle for Okinawa had been costly for the Marine

1st Division, with 7665 men killed, wounded, or missing. Japanese killed were estimated at 110,000, and tragically, it has been estimated that there were 150,000 civilian casualties—one-third to one-half of the island's prewar population.

Between Peleliu and Okinawa, 1st Division casualties (killed, wounded, and missing in action) totaled 14,191. Considering most came from three infantry regiments of a total of 9000 infantrymen, the losses for the two battles combined reached 157%. It is notable that, according to some records, there were more mental health issues from the Battle of Okinawa than any other Pacific battle. The causes were attributed to constant bombardment by Japanese artillery and mortars, unbearable conditions, a high casualty rate, and an enemy willing to fight to the death.

Some historians believe the large number of casualties from the Okinawan campaign led to the decision to bomb Hiroshima and Nagasaki, to avoid an Allied invasion of Japan. The Japanese formally surrendered on September 2, 1945, ending preparations for an American invasion.

After the Armistice, Marc's battalion headed to Tientsin, China, to ensure the surrender, de-arming, and removal of any remaining Japanese troops. (The battalion commander gave each of his officers a confiscated Japanese saber.) Having completed his two-year tour overseas in November 1945, Marc arrived in San Diego in December and within days started eastward with his seabag, looking forward to his honorable discharge.

Jaffe donated the saber to the American Legion in Williamstown, where it hangs prominently on their wall. *[Courtesy of the author]*

Reflecting on his long-ago service, Marc feels the best part was "being a Marine and being alongside some of the same men for years, even into China, and knowing you could depend upon each other." He doesn't mind talking about his military service; he thinks it relieves pressure in him and hopefully educates younger people to understand what the nation went through.

Marc Jaffe's experiences as a Marine Corps officer taught him that a leader must pull his unit together, provide examples and guidance, and energize his men for a difficult job, keeping everyone focused on the mission.

He cherishes the saying, "Once a Marine Always a Marine," feeling it applies to a special group of people and is a mantra not repeated by any other military branch.

Publishing Career

A little over a year after his discharge from the Marines in February 1946, and after exhausting his savings from overseas pay, an employment agency helped Marc find a job with the long-gone men's magazine *Argosy*. He spent a year as an editorial clerk, then joined the New American Library publishing company in October 1948 as Associate Editor with responsibility for Westerns and mysteries. He remembers his time at NAL as a great "wide-angle apprenticeship," learning the ins and outs of paperback publishing. It was the early days of the "paperback revolution," when paperbacks were distributed for the first time by magazine wholesalers and priced at 25 cents, and soon 50 cents. At the time, books on racks were available in cigar stores, drug stores, and supermarkets.

While editing a wide range of books, he specifically remembers working with Mickey Spillane, the crime novelist who invented the signature detective character Mike Hammer. Marc remembers working with Spillane on several of his early books and always thought he looked the part of a crime writer: a solidly built, rugged-looking

Irishman. Few people know that Spillane honed his writing skills by writing comic books.

Jaffe also remembers editing and even providing titles for three novels by Edgar Box, the pseudonym of the multi-talented young novelist, Gore Vidal, then much in need of financial support.

During this time, back in Philadelphia, both of Marc's parents were beginning to decline. Lily passed away, at home, from cancer in 1955. Later, in 1966, Samuel suffered the stroke that ended his life.

In 1961, in New York, during what many have described as the "golden age of paperback publishing," Marc joined Bantam Books as a Vice President and Editorial Director.

During Marc's years with Bantam, he had the opportunity to work with many well-known writers. He worked closely with Louis L'Amour, the famous spinner of Western tales. Marc, who had similar interests at an early age, got along well with him. L'Amour, a large, handsome man, looked and lived the part of a frontier man. Marc thought L'Amour was a great example of someone who decided he wanted to be a successful writer, and did it. He was hardworking, ambitious, and prolific at his craft, often writing three books a year for much of his career.

In later years, Marc became friends with Gore Vidal, who went on to write many well-regarded essay collections and more than two dozen novels, among them the widely acclaimed social satire *Myra Breckinridge,* which became a movie. Bantam published seven of Gore Vidal's books.

Of note, Marc was also friends with Norman Mailer, a class behind him at Harvard, who served in the Army during World War II. Mailer was famous for his debut novel, the highly successful *The Naked and the Dead*, and later became a public figure, even at one point running for mayor of New York.

While Marc was at Bantam, he was involved in creating Bantam Extras, also called "instant books." Bantam was one of the early companies to quickly publish a book, usually only days after an important current event, to capitalize on the likely reader's interest.

Over time, Bantam published over seventy "instant books" and was noted in the mid-sixties in the *Guinness Book of World Records* for the "fastest publishing."

Jaffe is often asked about the books he's published of which he is most proud. He has a ready answer: "I am certainly proud of the important novels for which I personally negotiated reprint rights, including Edgar Doctorow's *Ragtime* and William Styron's *Sophie's Choice*. But I am perhaps most proud of having initiated and supported an original series of foreign-language dictionaries — a project that had never before — or since — been attempted by a mass market publisher. After the initial titles had been published, I then personally arranged for reprint rights to a Latin-English dictionary and even a Hebrew-English dictionary in order to round out the entire series."

Marc also chanced to meet, in the late '60s, William Peter (Bill) Blatty, a successful screenwriter and author of several comic novels, at a New Year's Eve party where Blatty mentioned his difficulty in getting his publisher interested in a serious novel he wanted to write. Blatty, a devout Catholic, had learned of a priest in a middle-western city who had successfully performed an exorcism on a teenage boy. Blatty — probably wisely— decided to change the locale of the events to Georgetown and the sex of the main character to that of a young girl. Marc encouraged Blatty and told him he thought it would make a good story. Within a few days, Blatty and Jaffe made a modest deal, granting world publishing rights to Bantam. *The Exorcist* subsequently sold millions of copies and was adapted into a film, the screenplay of which, written by Blatty, received an Academy Award.

He also worked closely with historian Alvin Josephy, who advocated for Native American rights; William Stevenson, who wrote *A Man Called Intrepid*; and Poet Laureate James Dickey, for whom Marc edited his final novel, *To the White Sea*.

Marc's publishing career continued to thrive, and in 1980 he assumed a position with Random House. He became Random House Executive Vice President and Editorial Director of Ballantine Books, a Random House subsidiary.

Later in the 1980s, he joined the Houghton Mifflin Company, which announced that Marc would have his own imprint and focus on developing fiction and non-fiction titles.

As the years went on, Marc continued to edit (and freelances even today). His efforts include editing *Lewis and Clark—Through Indian Eyes* and *Latinos and the Nation's Future*. More recently he's worked with noted Berkshire author Kevin O'Hara on his fourth book, *Ins and Outs of a Locked Ward: My 30 Years as a Psychiatric Nurse*. Now, as a freelancer, he admits that he does "miss the opportunity of saying *yes,* the province of an editorial director."

Marc and his wife Vivienne have been married for over forty years and in 1989 moved from upstate New York to Williamstown, Massachusetts, with their family. Vivienne has authored four books, owned a business with Marc (Editorial Direction) in Berlin, New York, and has extensive editing experience.

Marc has four adult children, two from his first marriage, and used to play tennis several times a week. He enjoys reading books by Michael Connelly and Daniel Silva. He attributes his long life to a combination of genetics, luck, not being shot (in battle, or in all his years in New York City, or, jokingly, by his wife), and moderation in eating and drinking. This thin, five-foot-nine centenarian, with glasses, creased cheeks, and a white beard, still maintains his Marine Corps weight of 165, though he loves Italian and Mediterranean foods.

His advice on living long is something to take to heart: "You have to consider people and the world around you. You have to take responsibility for not only yourself but others. And read a lot."

Staff Sergeant John H. Quinn
Tail Gunner—35 Missions
WORLD WAR II

Jack Quinn had been with the 351st Bomber Group for several months. He had grown accustomed to his perilous role as a tail gunner on the B-17 bomber known as the "Flying Fortress." So far, his luck had held; not long ago, their plane had been raked by flak and lost two engines in an intense barrage over Germany. Their pilot, Lieutenant Barker, had been able to coast the wounded bomber back to France and crash land near Paris. The entire crew escaped from the shredded bomber and were able to return unscathed to England.

Just three weeks later, on September 28, 1944, the last mission of the month for his squadron (the 509th), Jack and his crew were bombing a critical synthetic oil refinery in Magdeburg, Germany, when a barrage of flak destroyed three of their four engines. They were hundreds of miles from their base in Polebrook, England. Lieutenant Barker, again piloting, was able to coax the shrapnel-riddled, smoking plane as far as Belgium, where the crew bailed out at the last minute. Most escaped injury, except Jack.

Early Life

JOHN H. QUINN, CALLED "Jack" by his family, was born on December 28, 1920, in Williamstown, Massachusetts, the son of Patrick J. and Mary Bridgman Quinn. The family lived on Glen Street, next to the Williams College campus.

Patrick and Mary were active in the local community. Patrick owned a dry goods store on Spring Street that was eventually converted to a shop serving Williams College students. Later in life, he worked as a New York Life insurance agent. Patrick also served in several civic positions for the town.

Jack's mom, Mary, a registered nurse, started the local Visiting Nurses Association and was its first district nurse working among the needy when trolley cars were her only means of transportation: "Fees in those days, Mrs. Quinn recently recalled, never ran over 50 cents, and she seldom made more than $10 in any month, even though as many as 192 calls were made in a single month" (*Transcript* 11-7-1952).

Tragedy struck the family when Patrick died unexpectedly at age forty-nine after a brief illness. Jack was only twelve years old.

After Patrick's death, to make ends meet, Mary would occasionally rent out Jack and his brother's bedroom to Smith College girls attending Williams College house parties on weekends, always making sure the girls were back by 10:30 p.m. Jack and his brother would sleep in the barn's loft behind the house, which they called their "sky palace." Jack also managed a paper route from an early age that helped support the family.

As a youngster, Jack and his chums would catch frogs on nearby ponds and sell them to the Greylock Hotel for five cents each. The chef then supplemented his menu with exquisite "French" frog legs. When he wasn't in school, Jack spent many a winter day cross-country skiing with his Irish setter, Suki. In the summers, he would caddy at the Taconic Golf course.

Jack graduated from Williamstown High School in 1940, where he played on its football team.

Some think that the solemnity of his dad's funeral left a lasting impression on Jack. It wasn't long after high school that he attended the New England Institute of Anatomy in Boston, graduating in 1941.

He briefly worked at the Graham Funeral Home in Springfield, Massachusetts, but a declaration of war with Japan upended his plans, and he enlisted on October 28, 1942, into the US Army Air Forces (USAAF), hoping to become a pilot.

After his basic training, Jack spent time at George Field in Illinois; Coe College in Cedar Rapids, Iowa; and at a field in Santa Ana, California, all pilot training areas. Due to an unforeseen illness and an overabundance of pilot applicants, he was placed into a B-17 gunner training program.

Extensive cross-training continued, and Jack spent another 15 weeks at an aircraft mechanic school in North Carolina. Then he was selected to spend time with a B-17 squadron in Miami Beach before returning to Coe College for more technical education.

The Plane

The Boeing B-17 earned its moniker when a news reporter covering its debut spotted the large number of machine guns sticking out of it and called it "a flying fortress." Boeing liked the name and quickly patented it.

Often described as a "lightweight aluminum cigar tube," the B-17 heavy bomber had a 100-foot wingspan and a ten-man crew crammed together in a long, narrow, claustrophobic space that smelled (at minimum) of sweat, cordite, cigarette smoke, urine, and dried blood. The plane used an interphone system for the crew to communicate. A narrow catwalk separated Jack, as gunner, in the back of the aircraft from the rest of his crew members in front.

The Fortress weighed in at 65,000 pounds, typically cruised at 150 miles per hour, and sported thirteen .50-caliber machine guns, eight of which were on moveable turrets with their feeder trays to provide 360-degree protection.

The B-17 had four engines, could carry 6000 pounds of bombs, and travel 2,000 miles on 2,800 gallons of gas. From their base in Polebrook, England, to Berlin was a little over 1000 miles roundtrip. Most importantly, the Fortress could fly and land with one engine.

B-17s flew in "combat box" formations. The layered V formations solidified their defenses allowing planes to protect neighboring B-17s from the swarms of enemy fighters. A German pilot once said that attacking a B-17 formation from behind was like trying to make love to a porcupine that is on fire.

The Fortress's crew of ten included a pilot, co-pilot, flight engineer/upper turret gunner, navigator, bombardier, radio operator, waist gunners (2), ball turret gunner, and a tail gunner. Crew members played multiple roles, often operating a machine gun in addition to their usual role as the plane drew close to its target.

The oxygen and electrical systems were critical to the B-17 crew members, especially concerning their ability to breathe at high altitudes and keep warm in sub-zero weather conditions.

As time drew close for Jack to be dispatched overseas, he was transferred to Las Vegas and Salt Lake City in the spring of 1944, where he attended armament and gunnery schools. Learning to field strip and reassemble a .50-caliber machine gun while blindfolded (with gloves on) was one thing, but gunnery school also imparted to Jack the very serious necessity of rapidly identifying friendly versus enemy aircraft.

The gunners were taught other crew members' responsibilities, in order to quickly take their place if (when) they were wounded. Classes also included radio operation, mechanics of the B-17 bomb racks, instructions on the plane's electrical and oxygen systems, and basic first aid. At 30,000 feet, everyone was a medic.

All classes included shooting skeet and learning the principles of "leading" a target. Soon the gunners would be in an environment where they would be leading enemy targets traveling hundreds of miles an hour directly at them.

The following news article appeared in *The North Adams Transcript* on March 3, 1944:

Will be Assigned to a Flying Fortress—Corp. John H. Quinn Wins Rating as Air Force Gunner

Corp. John H. Quinn of the Army Air Forces, son of Mrs. P.J. Quinn of North Adams, is visiting here on a short leave from Las Vegas, where he has won a gunners rating. Corp. Quinn will report to Salt Lake City, where he will be assigned to a Flying Fortress as a top turret gunner."

Several months later, in the summer of 1944, Jack headed overseas. It is believed he sailed, as did most Air Force crew members, on a troop convoy out of New York, and after five days aboard the ship, he disembarked in Scotland and completed his journey to his base in England by train and a 4 X 4 truck.

Jack was assigned to the 8th Air Force, which had over forty bases in England, all focused on defeating Nazi Germany. He was sent to the 351st Bombardment Group, located at Polebrook, England, in an area called "the Midlands," over a hundred miles inland from the North Sea. Before its conversion to an Air Base, Polebrook had been part of the Rothschild estate, formerly a potato patch.

The 351st had been at Polebrook for over a year, and its four squadrons of B-17s (508th, 509th, 510th, and 511th) had seen continuous action. Each of the four squadrons had twelve planes, and between repairs due to enemy action, it was hoped that they could keep six or more B-17s flyable at any one time.

John Quinn as a young crew member. *[Courtesy of Quinn family]*

When Jack arrived, he was assigned as a tail gunner on the ship *Star Duster* in squadron 509, and his missions began almost immediately.

He settled into his new home away from home, a silver-ribbed, uninsulated Quonset hut bolted to a concrete floor that he would share with his crew and twenty other men from two different crews. The half-moon building was heated with a central potbellied stove that probably used coal and was lighted with a few low-wattage bulbs dropped from the ceiling. He had a single metal bed with a thin, canvas-covered straw mattress and an itchy, woolen army blanket. The latrine and wash-up room were in a separate building.

During their brief orientation, the new men learned that most of the squadron's strategic targets would be in Germany, including ball-bearing plants, locomotive and tank factories, oil refineries, and anything that could support the Nazi war effort. The 509th would also target sites in France and Belgium, trying to cripple submarine installations, V-1 rocket launching sites, and harbor facilities.

British and American leaders had reached an agreement in January 1943 that they would engage in a combined bomber offensive: The British would continue to bomb at night and the Americans would bomb by day. The British had wanted Americans to join them in nighttime forays, but Army Air Force Generals thought that would not be as effective as around-the-clock bombing.

While the aircrews were sleeping, mechanics hauled bombs to the planes, hand-winched them aboard, hung them on racks in the bomb bay, and inserted fuses. The mechanics would then insert cotter pins to prevent the tail from spinning and arming the bomb. It was dangerous and delicate work. Then mechanics inspected the entire plane's engines, hydraulics, brakes, and electrical and oxygen system.

At the same time, the armaments group hauled boxes of .50-caliber ammunition onto the plane, placing boxes at each gunner's station. Then they made sure each turret worked.

Jack and his fellow crew members were awakened next, usually between 1:00 and 3:00 a.m., had breakfast, and then headed to the briefing room. There, a large map of Europe would be uncovered, and their main target areas were assigned, along with secondary and tertiary targets should the main one be overcast or unbombable. Places with heavy flak or possible enemy fighter concentrations were noted.

Cultural buildings such as cathedrals and opera houses were sometimes noted as being off-limits to bombing.

The main foes of the B-17 were German fighters, specifically the Messerschmitt Me-109 and the Focke-Wulf Fw-190, and particularly dangerous was flak (shrapnel) from German anti-aircraft guns called 88s.

The Me-109 was a defensive fighter with more aerial kills than any other aircraft in World War II. The Fw-109 was considered a more devastating plane against Allied bomber formations due to its speed and ability to climb.

Eighty-Eights (with a caliber of 88mm) were highly accurate cannons that, when fired vertically, could reach heights of almost 30,000 feet. The cannons produced what was known as "flak," a contraction of the German word *Fliegerabwehrkanone*. A direct hit by one of their 20-pound high-explosive shells could destroy a B-17 bomber. Near misses could shred the plane with hundreds of pieces of shrapnel, easily knocking out an engine and seriously wounding crew members.

These ten-man crew-served weapons could fire 20 rounds a minute and were usually within a battery of four weapons directed by fire control computers.

Aircrews could see and feel the exploding angry black clouds of flak explosions, never knowing when the next one would hit their ship. Though the B-17 was considered one of the United States' most durable aircraft, it was made out of aluminum, and a person could easily punch a hole through the fuselage with a screwdriver. By war's end, flak would bring down over 5000 Allied planes . . . enemy fighters would destroy around 4000.

Once the briefing was complete, Jeeps would take gunners to the armory to collect their guns, where they had been cleaned and oiled, and then drive to their planes. Jack would crawl to his narrow tail-gun position to install his two guns, wrestling the 80-pound guns into their yokes.

The pilots would be visually inspecting the plane with the ground chief.

The crew would dress in their fleece-lined flight clothing, boots, gloves, parachute harness, and life preservers, and as they approached higher altitudes, make sure their electrical suits and oxygen masks were in working order. Body armor was worn when they entered enemy air space.

Once the pilots started their planes, the danger began immediately as they took off in the darkness. With many other airfields nearby, groups would corkscrew up, directly over their own airfield, to avoid incurring collisions. Once they reached an altitude of 16,000 feet, they'd level off, join other squadrons, and head toward their target.

Gunners would fire and clear their guns over the North Sea to make sure they were operable, and then came the humdrum flight into Germany, which at one hundred and fifty miles per hour would take two to three hours before high alertness was called for. The close formations flew level and bombed at heights ranging from 20,000 to 30,000 feet, where temperatures reached 40 to 50 degrees below zero.

The bombers had to fly straight and level for the Norden bombsight to work effectively, making the planes predictable and vulnerable to the 88s that were staged in batteries around strategic targets.

Jack was all alone as a tail gunner in the furthest reaches of the plane. Everyone else had someone nearby to talk to. Although all crew members were connected by interphone, he was by himself, and the engine noise blocked most of the conversation with him over the phone.

Alert facing outward, protected by a bulletproof cupola and some armor plating, he would man two .50-caliber machine guns. Perched on his knees, sitting on an oversized, bicycle-like seat that would support part of his weight, he would be ready to intercept enemy fighters approaching the rear of the B-17. Sitting/kneeling in his narrow firing position in the plane's tail was difficult, and his legs would cramp up. He sometimes needed to turn around, reposition himself, and stretch out.

At a certain altitude, Jack would begin breathing from his oxygen mask and always welcomed the warmth provided by his heated, fleece-lined body suit connected to the plane's electrical system. His parachute was hung on a nearby hook readily available.

His first missions are believed to have occurred in Germany, and the first eventful one occurred on August 8, 1944, when the *Star Duster* experienced heavy enemy flak over Ludwigshafen, while it was bombing an oil refinery. With two badly damaged engines, the pilot coasted the plane back to occupied France and crash landed. The crew members, traveling separately and disguised, made their way to Paris, arriving just days before it was liberated from the Nazis. They were hidden in a wine cellar by the French resistance. They had little food but plenty of wine and stayed hidden until the city was liberated, then resistance members spirited the aircrew back to their base in England.

Less than one month later, on September 28, 1944, while bombing a synthetic oil refinery in Magdeburg, Germany, flak hit Jack's B-17 after it dropped its bombs, knocking out three engines.

Another harrowing close call from which they barely walked away. *[Courtesy of Quinn family]*

The pilot coaxed the smoking plane for miles until the crew could bail out over Belgium near the town of Ypres. Parachuting from less than 1000 feet high and struggling with a chute that would not immediately open, Jack landed hard, breaking his ankle. The rest of the crew was more fortunate, landing in hay mounds, canals, or softer ground. Local partisans intercepted the crew, splinted Jack's ankle, and provided them underground assistance for several days until they were whisked back to England, where Jack spent weeks recovering at the Army hospital in Peterborough.

Quinn is third from the left, after what was comparatively a safe landing back in England. Note their broken propeller. *[Courtesy of Quinn family]*

Once recovered, sometime later, the crew was again forced down in Belgium, crash landing behind enemy lines. When approached by Belgian farmers, a German-speaking crew member started speaking in German to calm the gun-toting residents. Realizing the farmer's animosity toward Germans (and that the crew was about to be shot), they quickly yelled "Americans! Americans!" They were greeted with smiles and turned over to partisans who helped them return to England.

Thoroughly exhausted crews who returned from their six- to eight-hour harrowing missions were extensively debriefed, given a shot of whiskey, and sent to bed. They usually went to sleep only 24 hours before their next mission.

Jack's crew had been extremely fortunate, crashing behind enemy lines three times and quickly returning to friendly territory. Aircrews captured over Germany were often mistreated and tortured, and as the war intensified, many were executed.

Aside from air combat, where the odds of a B-17 crew member surviving the war were only one in four, there were other dangers. Flying in an uninsulated and unpressurized environment required electrical suits, gloves, and boots, and if electrical contacts were broken through shrapnel, frostbite was very common, and amputations resulted.

Clogged oxygen systems (due to air sickness) could freeze hoses, causing men to pass out at 25,000 feet and, becoming unaware of their situation after two minutes, die from asphyxiation. Oxygen checks were done every few minutes over the interphone to prevent this. Clearing clogged guns at high altitudes with bare hands resulted in crew members' skin sticking to the metal and ripping off in strips.

Studies at the time indicated a certain amount of paranoia and anxiety associated with the men's missions, and this began to accumulate, yielding a feeling of helplessness, never knowing when the next barrage of flak would take down a ship.

Jack's B-17 crew continued to fly missions throughout Germany, France, and Belgium and supported the Army during the Battle of Normandy and the counterattack that occurred during the Battle of the Bulge in December/January 1944–45. In early spring 1945, Jack reached the haloed number of 35 combat missions and was rotated off combat duty.

Exhausted by his ten months of combat air service, Jack left England by troop ship, reentered the United States through New York City, and was involved in an honor parade before returning to Fort Devens to be honorably discharged on June 11, 1945. The war in Europe ended three months later, in September 1945.

During World War II, half of all the Army Air Force casualties were with the 8th Air Force. Over 26,000 men would die, and 4754 B-17 Bombers would be shot down. In Jack's unit (just one squadron of many), 175 B-17s and their crews were lost.

Quinn's *Star Duster* jacket, with its 35 bombs signifying 35 missions, worn in this photo by his son Patrick. *[Courtesy of Quinn family, Mike Remillard Photos]*

An interesting postscript: The most notable crew member of the 351st was Clark Gable, the famous actor. Gable, wanting to be part of the war effort, flew a number of missions as a gunner and photographer out of Polebrook. One of his missions turned out to be a close call when the heel of his boot was shot off, and flak narrowly missed his head. Gable also suffered frostbite for not wearing his heated gloves. When Hitler heard of his exploits, he quickly offered a reward for Gable's capture.

After his discharge, Jack returned to work as a funeral director. During one of Jack's furloughs, he had been introduced through a relative and encouraged by his mom to meet with a "very nice teacher" they knew. Jack asked Lois Boland, a recent graduate of Our Lady of Elms College, on a date, and they clicked. They were married in 1948.

In 1949, Jack opened the John H. Quinn Funeral Home and would provide empathetic service for over thirty years. (At one point he ran another funeral home in North Adams.) Never one to remain idle, he was also involved in town affairs and served for years as Town Clerk and Auditor. During this time, Jack and Lois had three sons and two daughters.

Lois herself was busy working for three decades as the first female Berkshire County juvenile probation officer. She also was involved in working with adoptive families, was a founding member of the local senior center, and was voted Williamstown's "Mother of the Year" in 1971.

Jack and Lois ran a successful wholesale dairy products business, selling to restaurants and dairy bars for years. Jack also worked third shift at a local cable company to ensure his work hours wouldn't interfere with the funeral business. His sons still remember being called out late at night to pick up a recently "passed" person and bring them to the funeral home in their dad's absence.

Jack said little about his war experiences until he was in his seventies. His son remembers watching a World War II movie with his dad that featured anti-aircraft fire "flak" and his dad commenting how awful flak was and how it could easily knock a plane out of the sky.

Jack also told his son that before leaving for the war, knowing how dangerous a gunner's position was, he sold his Chrysler Woody convertible and gave the money to his mother because he never expected to come back from the war.

One time, the family visited an air museum with a B-17, and everyone marveled at how Jack ever got out of the small hatchway not once but three times while his plane was on fire. Jack explained that being a tail gunner was such a tight location at the back of the plane, and the action so intense, one needed always to be critically aware of his surroundings. He could remember on any number of occasions not having enough fuel to complete their mission and having to coast their way back to the base.

Jack's son recalls, "After being shot down, though he loved airplanes, he was fearful of flying again. The only times he flew were [once] when he and my mother were wintering in Florida and [friends] invited them to spend a week in Jamaica. Second was when they flew to England when my brother Dan was at the London School of Economics. They toured England, Scotland, and Ireland—that was it for Jack to fly again. I really can't blame him; I think I would be flight shy as well."

Jack's service to his country was described in a *Transcript* article dated March 8, 1945, when only part of his overseas tour had been completed. The article noted that he was a "Veteran of aerial assaults on every major industrial city in Germany, including Berlin, Kiel, Munich, Cologne, and Frankfurt." He was awarded an Air Medal with 5 Oak Leaf clusters, a Purple Heart, the European Theater Medal, an American Victory Medal, and two Distinguished Unit Citations. In ten fearsome months, over thirty-five sorties, young John H. Quinn showed himself to be a man of undaunting courage and patriotism.

Jack and Lois, who vacationed in the later years of their lives part-time in Florida, were married for over fifty years. Jack passed away about ten years after Lois. It can be said with certainty that Jack's long life was a remarkable gift.

Captain Robert J. Guillotte & Yeoman 2nd Class George L. Guillotte

Fighter Pilot & Submariner

WORLD WAR II

Two brothers, born just a year apart, both joined the fight against fascism in 1940. One joined the Army Air Corps and eventually became a fighter pilot; the other enlisted in the Navy and entered the submarine service. Their deployments intersected during the attack on Pearl Harbor and, together, using an ambulance, they whisked numerous critically wounded servicemen to the hospital.

Shortly afterward, separated again, the older brother became a highly decorated fighter pilot, shot down over France and captured after a three-week perilous effort trying to escape the German Gestapo. His younger brother, now several oceans and thousands of miles away on a submarine, was avoiding enemy destroyers and helping to sink Japanese merchant ships.

Caught in the middle of a worldwide conflagration, both young men wondered if they would ever again see their beloved Williamstown.

ROBERT WAS BORN FIRST, on February 10, 1922, then on February 23, 1923, his brother George arrived. Both were born in Springfield, Massachusetts, to George D. and Laurette (Roberts) Guillotte.

The boys' father, George, a French Canadian, was an excellent carpenter and was in training to be an architect. Laurette was a homemaker, and her family name, Roberts, was well known in the area as the grocers behind Roberts Fine Foods.

Baby George on the left, Robert on the right, in one of the few early family photos. *[Courtesy of Guillotte family]*

In 1926, in a startling, tragic event while at a church picnic, the senior George's canoe was hit by lightning, and he and a cousin were electrocuted. The bodies were not immediately recovered. The story recounted by family over the years tells of a priest who was called to the scene. He floated a host (the wafer or bread used in Communion) on the water, and it stopped directly over a spot where the searchers had not looked. From that point, the bodies were found deep in the weeds.

Understandably, the gruesome accident devastated Laurette, who'd birthed a third son that same year, joining the very young George and Robert. The Roberts family helped her for a time, and then in 1934, Laurette married Raymond Boussy, an insurance salesman whose territory included northern Berkshire County. The family moved to Williamstown.

The boys attended local schools, George one year behind Robert. Shortly after arriving in Williamstown, the boys participated in a Parent Teacher Minstrel Program and sang a duet together. Otherwise, both engaged in different activities. Robert had a radio debut in a school play, competed in a statewide poster contest, and, most of all, was known for his math skills. George was active in many class programs and pageants and supported a print exhibit at the school.

It is thought that the boys worked some of their summers at Roberts Fine Foods in West Springfield with their grandfather and uncles.

It appears that Robert transferred high schools in his senior year, graduating from West Springfield High School in 1939, although he maintained his residency in Williamstown. Robert had high math acuity and may have switched schools to find more challenging courses. It is reputed that at a very young age, he'd enjoyed mature, encouraging discussions on the subject of the trisection of angles with a Williams College math professor.

In March 1940, nine months after graduation, Robert enlisted with the Army quartermaster's corps. He was the first male in Williamstown to enlist for service in the military in World War II. After his basic training, Robert was transferred to Hickam Field in Hawaii.

George spent some time with the Civilian Conservation Corps in Connecticut, then decided to follow his older brother into service. He left school early, enlisting in the Navy on December 6, 1940. After basic training at Naval Station Great Lakes in Illinois and then some weeks at Yeoman's School in Rhode Island, he reported to the USS *Pelias*, a submarine tender, on September 5, 1941. He participated in its "shakedown cruise" before leaving San Diego for Pearl Harbor, arriving in late November 1941. As a Yeoman, George would work directly for the ship's executive officers. This highly confidential position managed the ship's messaging, correspondence, record-keeping, and files.

The morning of December 7 found Robert in a hospital bed at Hickam Field with an infected leg, and George on the USS *Pelias* moored at the submarine base at Pearl Harbor.

Then the Japanese planes arrived, beginning their bombing and strafing runs on the American base. It was relentless. They passed about 100 yards off the left side of the *Pelias*. The crew, called to their stations, quickly reacted and were able to knock down one enemy torpedo bomber and damage a second one.

During the attack, Private Robert Guillotte abandoned his bed at the hospital, commandeered an ambulance, and began picking up the wounded, scattered as they were around the battered base. At the same time, the captain of the USS *Pelias* released some sailors, among them George, to help with the numerous casualties on shore. Somewhere in this confusion, the brothers met and began working together, transporting casualties to the base hospital. When Robert returned to the bombed hospital, he found that his bed had been twisted around roof support, and some patients nearby where he had lain had been killed.

Because of his initiative, Robert survived, and he also received a citation from his commanding officer.

Several weeks later, the brothers spent Christmas together and then said goodbye. They were not to see each other again for years. Their paths would be separated by oceans and thousands of miles, fraught with dangerous situations.

Robert remained in Hawaii for a short time, attending the West Point Preparatory School and passing an exam that would allow him to become an aviation cadet, a long-sought-after dream. After qualifying for pilot training, Sgt. Guillotte returned to the United States in March 1942 and passed a second test that enabled him to skip the first training phase.

He started flight training in Santa Ana, California, and finished his final six months at Luke Field, Arizona, where he received his wings. After additional instruction, he was commissioned in May 1943.

Robert had been trained to fly a P-47 Thunderbolt, a rugged fighter bomber known as "the Jug." (It is thought that the nickname was most likely derived from the fact that if the plane stood on its nose, it would look like an old-fashioned glass milk bottle.) The P-47 had an eighteen-cylinder power plant, carried 1000-pound bombs under its wings, and sported eight .50-caliber machine guns. Its pilots often referred to it as the "flying tank."

Robert Guillotte, getting in some flight time with the P-47. *[Photo by J.J. Lagana;* The History of the Hell Hawks *by Charles R. Johnson]*

The P-47 soon became well known for providing devastating close air support for Allied ground troops. Its incendiary, armor-piercing .50-caliber ammunition was acclaimed for destroying formations of German armored vehicles or troops in the open. The P-47 was often referred to by the American infantry as their "flying artillery."

While Robert had been learning to fly the P-47, in early 1942 George was redeployed aboard the USS *Pelias* with its 900-man crew, part of the Asiatic fleet, to Fremantle, Australia, near Perth. There they had the critical job of repairing and refitting scores of submarines. The USS *Pelias* served a crucial function in her sub-tending duties and at wars' end would receive a battle star for her service.

It was a time of adjustment for the submarine service, after narrowly surviving the Japanese assault on Pearl Harbor. Shortly after the attack, the President of the United States declared "unrestricted warfare" on any ship flying the Japanese flag. This allowed the targeting of all vessels, including warships, troopships, and merchant vessels. Initially, the US submarine fleet was ineffective in taking advantage of this unrestricted status because of inadequate training, poor tactics, lack of aggressiveness, and defective torpedoes.

Over time, these problems were resolved, and the submarine fleet became a critical factor in denying resupply to the islands that Japan had occupied and even to Japan itself. Submarines became a major factor in collapsing Japan's economy and ending the war. The service was credited with sinking 1300 merchant ships, approximately 200 warships, and over forty troop transports that carried thousands of island reinforcements.

In June 1943, George was transferred to the USS *Chanticleer*, a Submarine Rescue Vessel supporting submarines working outside Australia. The one hundred sailors on the *Chanticleer* would train divers, carry out salvage operations, and refit subs between patrols. In one case, it towed a captured Chinese junk called a *bandoeng* to Fremantle, Australia, where it was handed over to Intelligence agencies. (A *bandoeng* was a cleverly disguised 80-ton fishing junk with sails that could attain 10 knots by using its diesel engine and was used by special ops for reconnaissance missions.) After delivering the *bandoeng*, the *Chanticleer* worked in the Philippines to locate and salvage sunken US and Japanese vessels creating hazards in the waterways.

While George was stationed on the USS *Chanticleer*, his brother Robert, now Lieutenant Robert Joseph Guillotte, married Virginia Lee Maxson, after which they resided at Camp Springs, Maryland, for several weeks. Then in July 1943, Robert received orders to return overseas. He was assigned to the 365th Fighter Group and left for England on the *Queen Elizabeth* in December 1943 as part of the Ninth Air Force.

The 365th Group, nicknamed the *Hell Hawks*, was initially based at RAF Gosfield, eventually moving into Quonset huts in Beaulieu, England. The group spent January 1943 adjusting to their new surroundings and flying tactical formations; in February, they began their first combat missions.

The daily pattern remained the same: very early breakfasts, followed by target intelligence briefings, then pilots were Jeeped to their planes for checks and take-offs. Each one-man aircraft would circle above the base, join its squadron's formation, and head to the assigned mission. In February and March, *Hell Hawks* focused on escorting bombers to their missions, often in northern France.

In March, Robert was part of a group of P-47s credited with shooting down six German Focke-Wulf 190 fighters and probably two others after an extended aerial battle.

In April and May 1944, with the approach of D-Day, the *Hell Hawks'* mission changed from escorting other bombers to, themselves, dive bombing. The destruction of any transport that would interrupt reinforcement and weaken the German response to any planned Allied invasion in the summer of 1944 was considered a likely target. The unit began bombing rail lines, railyards, locomotives, troop trains, and bridges.

In a special section of *The North Adams Transcript* dated May 3, 1944, the Ninth Air Force credited Lieutenant Robert Guillotte, after resolving a stuck bomb release, with unloading his bomb "squarely on railroad tracks" that resulted in a "whole Nazi troop train falling into Lieut. Guillotte's Crater."

Unfortunately, Robert's good fortune ran out several days later when he was shot down on a bombing run on a steel bridge at Vernon, France. On May 7, 1944, Robert and another pilot, flying at very low altitude, directly attacked the bridge, and both came under withering fire from nearby flak towers. The pilots immediately felt the shuddering impacts as the cannon fire struck their planes. Robert still had the presence of mind to drop both of his thousand-pound bombs directly on the target even while the cockpit of his plane (the cleverly named *Guillotine*) filled with smoke. His plane was fully on fire as he bailed out just four miles from the target after only climbing several hundred feet. After deploying his parachute so close to the ground, his wiry, 145-pound frame suffered a vicious impact, but he survived and quickly left the crash site.

It's worth noting that while declassified records indicate the government believed he may have survived, they state: "Extent of search: None—Incident occurred too deep in dangerous enemy territory to make searching practical."

No one was coming to help. For the next three weeks, Lieutenant Guillotte was on the run from the Germans, with the ultimate hope of reaching Spain. He received civilian clothes from some local residents to wear over his uniform, although most remained highly suspicious, suspecting he might be a Gestapo agent. For the most part, Robert stayed on the move, traveling alone and avoiding German patrols, sentries, and aerodromes. Occasionally he swam through rivers to bypass long convoys of trucks.

As he closed in on France's border with Spain, he was aided by the French Underground movement. A young woman accompanied him on a train to give him the appearance of being part of a couple. Robert, who spoke no French, played a deaf-mute when asked for his ticket. The conductor and a nearby Gestapo agent became suspicious, and he and his "partner" were strip-searched in a nearby train car and immediately arrested when they saw his uniform under his civilian clothes.

Taken to Gestapo headquarters in Bordeaux, they were interrogated at length. The commander wanted the names of all the people who had helped Robert, and he refused. At one point, the commander threatened Robert's and the young girl's life. Robert still refused. Abruptly, the Commander opened his balcony doors, and directed a firing squad below to shoot the young girl. Within seconds she lay dead.

Robert was shocked but still refused; handcuffed, he was led away to prison, wondering what would happen to him.

For the next few months, he would pass through several prisons and undergo many interrogations, then in July 1944 he was herded onto a train with other English and American airmen and sent to

Stalag-Luft 3 in Sagan, Poland. Robert would soon learn that the POWs called themselves "Kriegies," short for *Kriegsgefangenen*, the German word for prisoner of war.

One of the ironies of war occurred amid the severe camp privations, hunger, and lack of medical treatment; the Swiss Legation notified camp authorities that Robert had been awarded the Distinguished Flying Cross for his heroic efforts in bombing the bridge in France. On a Saturday, hundreds of German and American troops stood at attention while Robert (and others) received medals for actions against the Germans.

The letter from the delegation reads as follows:

> *To the Senior Officer—Stalag Luft 3.*
>
> *Dear Sir,*
>
> *On behalf of the American Government, we have the pleasure of forwarding you a list of names of American prisoners of war who have been cited for awards. We would be glad if you would inform these prisoners accordingly and convey to them our sincere congratulations.*
>
> *We have the honour to be,*
>
> *Yours very truly,*
> */s/Swiss LEGATION*

The individual award and citation read:

> *1st Lieutenant Robert J. Guillotte, P.O.W. No. 6658*
>
> *DISTINGUISHED FLYING CROSS*
>
> *"The Distinguished Flying Cross is awarded for accomplishing a low-level dive-bombing attack on the railroad bridge at Vernon on May 7, 1944."*

In January 1945, in a haste to avoid the fast-approaching Russian troops, Stalag-3 POWs were quickly evacuated on foot in the bitter cold. For weeks with little food, rest, or cover, they were forced to march under the most adverse conditions and, at one point, packed into cattle cars before arriving at Moosburg, Germany, and Stalag 7A, their home for the rest of the war.

There would be about 1800 Americans, among them Lieutenant Guillotte. Records indicate that upon arrival everyone was shaved from head to toe; so many had not bathed in months that lice infestation was rampant among the men. The trip had added to the men's woes: dysentery was common throughout the entire group, and with no heat and very little food, the new Stalag offered little solace.

George in the foreground, part of a work crew. Note the designation 391 for the USS Pomfret. *[Courtesy of Guillotte family]*

While Robert was in Stalag 7A, George had been promoted to 2nd Class Yeoman and, late in the war, joined the USS *Pomfret* (391), a Balao-class submarine on her last war patrols. It was an assignment George volunteered for. He had always wanted to be part of the undersea warfare group and was delighted to finally go beneath the waves. He handled all the boat's administrative work.

The *Pomfret* was a diesel-driven submarine that propelled at a submerged speed of 20 knots. The boat was 311 feet long and held a crew of 80 men within that confined space. Its armament included six forward and four aft torpedo tubes and several deck guns. It operated mainly out of the East China Sea, searching for and sinking small crafts and ships and destroying enemy mines.

In his capacity as yeoman, George was given an additional duty that would haunt him for the rest of his life. The captain ordered him to take pictures of all the vessels that the *Pomfret* was responsible for sinking. So, once the enemy vessel had been disabled or was breaking apart, the *Pomfret* would surface, and George was required to take pictures of the disintegrating ship, which included its dead, drowning, and desperate passengers. The actions were within the bounds of unrestricted warfare, but the memories could never be forgotten. Later in life, George shared them with his son.

This portrait of George Guillotte can be seen today at the Williamstown Historical Museum. *[Courtesy of Guillotte family]*

After her wartime patrols, the sub was homeported in Hawaii, and George served aboard her during trips to the Philippines and China. As a notable aside, Jimmy Carter, the future President of the United States, served aboard the *Pomfret* from 1948 to 1951. He is the only United States President to date to qualify as a submariner.

For her World War II service, the *Pomfret* was awarded the Asiatic-Pacific Campaign Medal with five battle stars, the World War II Victory Medal, The Navy Occupation Medal, and the China Service Medal.

In the meantime, the Allies were getting closer to Moosburg and freeing POWs along the way. For some, it would be too late. Robert remained alive and incarcerated, but in late April, the prisoners noticed that the German guards had left. Not long afterward, they watched in amazement as General Patton entered the Stalag in his Jeep. Immediately, arrangements were made for mobile kitchens and medical assistance.

After a brief recovery, the men were flown out of Germany to France from a nearby airfield. It was May 7, 1945, the date of Germany's unconditional surrender. The men would make several stops before reaching home. They would be trucked to Le Havre, France; sail on a transport to the United States; and travel to Camp Kilmer, New Jersey, for processing.

Robert convalesced for several months in Florida, recovering from the deprivations of being a POW. Then he returned to Virginia to be with his family. Robert officially left the military as a Captain on June 26, 1946, resigning from the 149th Fighter Group Squadron, Virginia National Guard.

Robert's medals included the Distinguished Flying Cross (two clusters), the Purple Heart, the Air Medal with Clusters, the American Campaign Medal, the European–African–Middle Eastern Campaign Medal, and the World War II Victory Medal.

At war's end, George was transferred to the USS *Plaice* (390), another Balao-class submarine, and he remained with her, operating in the Pacific. (The USS *Plaice* also had a highly aggressive and successful record against the enemy during the War.) George returned to Mare Island, California, in late 1946/early 1947 to be discharged from the Navy.

George would be awarded the Asiatic-Pacific Campaign Medal with five battle stars and the World War II Victory Medal. His seven years of service included three years in Australia and two years in China, Japan, and the Philippines.

After George's honorable discharge from the Navy, he returned to Williamstown and decided to attend the Henry Ford Mechanics School in Dearborn, Michigan, to learn a trade. As a yeoman, he was a great typist, but was concerned that it would not be a helpful civilian occupation.

Around the same time in 1948, he married Margaret Mary Hanlon, a nurse who was a local girl from Williamstown, and together they would raise five children.

George did not find his trade school immediately useful and so he returned to finish high school, becoming the sole proud veteran to graduate and receive a Lions Club scholarship with the Williamstown High School Class of 1950.

For several years, he worked for local merchants in downtown Williamstown, after which he became an insurance salesman for several notable companies, among them Boston Mutual, Prudential, and Metropolitan Life. For the following twenty years, he enjoyed owning and operating a service station, George's Gulf Station, in Williamstown. At one point, the Gulf Corporation recognized his outstanding customer service and awarded him a new Ford LTD convertible.

When not working, George was often spotted transporting his kids to and from school in a large but old passenger station wagon. For fun, he loved square dancing with Margaret and also had a hobby trying out new cameras. George passed away in 1996.

After Robert's convalescence leave, he returned to Virginia to rejoin his wife and daughter, and their family grew to four children. Following his military service, Robert worked for over twenty years for the National Advisory Committee for Aeronautics (NACA, now NASA) at Hampton Roads, Virginia, as an engineer in dynamic analysis. He was one of the lead programmers working on the Mars Viking Lander.

Robert loved traveling, boating, and flying. After retirement, he opened a workshop and called it Ship'n'Shore. He produced beautiful woodwork for boats, homes, and offices. He built grandfather clocks for all his family. He also loved golf, bowling, working on his computer, and visiting family and friends. Robert passed away ten years after his brother, in 2006.

The good fortune of George and Robert can, in retrospect, seem almost providential. Two brothers just a year apart in age from the town of Williamstown were sent all across the planet in their country's broad efforts to stem a global conflagration. They lived through a chance meeting in Pearl Harbor on the very day it was attacked, together proved their heroism in the face of grave destruction, then were dispersed again thousands of miles apart. Both endured at times desperate situations, but both survived, came home to a grateful country, and carved out a good life for themselves.

It was almost, but not quite, as if they'd never left.

PILGRIMAGES

VISITING THE PHYSICAL LOCATIONS where some of these patriots lived, or the cemeteries where they now rest, became a personal pilgrimage for me and was very meaningful. All of the following destinations are free and open to the public. I invite and encourage you to visit them to participate in the history all around you.

I remember lingering by the home of **Rudy Konieczny** at the foot of Mount Greylock and envisioning him skiing down the mountain right into his backyard. A short time later, I visited his grave at Bellevue Cemetery in the veterans' section where it had a view of that same mountain and the trails he conquered so many times.

On another day, I stopped by Maple Street Cemetery where **Reuben Whipple**, **John Welch**, and **Sterling Burnette** are buried. Sterling's grave marker notes he was "killed in France, November 11, 1944." His wife Veronique, who never remarried, joined him 64 years later in 2008. John Welch's headstone is located in a peaceful spot very close to one of the few remaining Quaker Meetinghouses in the United States; his USS *Brooklyn* flag, on the second floor of the Adams Free Library, is also worth a visit. Above the Rail Trail in Adams there is a bridge dedicated to Sterling Burnette and Walter Bednarz.

I stopped at the two squares in town that Adams dedicated to its patriots: Rusek/Whipple Square and Caron Square. Rusek/Whipple Square sits at the convergence of Crandal, Orchard, and Center Streets leading to different areas of the town. The Caron Square tribute, just yards from the busy Adams Post Office, is appreciated daily by hundreds of residents.

Daniel Petithory's grave in the Cheshire Cemetery is well tended by his family. Not far away in Cheshire is the towering, rock-studded

Sterling Burnette is honored on a bridge above the Adams Rail Trail. *[Courtesy of Eugene Michalenko]*

Memorial Day in Adams, at Rusek/Whipple Square.*[Courtesy of the author]*

monument to **Joab Stafford** standing lonely in the middle of a field. It gave me an appreciation of his last-minute gathering of a group of men, marching off to face unknown forces in a desperate fight. They may have known they'd be walking thirty-plus miles on mountain trails and muddy, rugged roads to protect our liberty, but couldn't know they'd change the course of a war.

Traveling to North Adams, I stopped by the local skating rink named after a Vietnam War hero, **Peter W. Foote**, who rests nearby in Southview Cemetery not far from his namesake building. A number of other North Adams patriots also rest in Southview: **Michael Cirullo**, **Michael Scarpitto**, and **Michael DeMarsico**, who also has a monument at the nearby North Adams Armory.

I visited the fascinating Soldiers' Circle at Hillside Cemetery in North Adams where **Elisha Nims**, slain at Fort Massachusetts during the French and Indian War, lies among friends. Arranged literally in

The Soldiers' Circle at Hillside Cemetery. *[Courtesy of the author]*

a defensive circle atop a knoll, these former soldiers keep vigil over the now-closed sprawling cemetery . . . eternally on watch.

Not more than 100 yards away, **John E. Atwood** rests peacefully, probably the only Berkshire resident to have heard Lincoln's Gettysburg address.

Visitors to St. Joseph's Cemetery in North Adams will find the final resting places of **Dr. George Curran** and **Henry F. Caron**. (The younger Caron, **Henry L.**, is buried at the National Cemetery in Bushnell, Florida.)

Just down the road in Williamstown, Thompson Memorial Chapel contains a marble dedication plaque to **Ephraim Williams Jr.** on the first floor and his vault on the lower level. Williams Library (Special

Collections) located behind the Chapel has several of his relics available for viewing, along with personal effects of alumnus **Charles W. Whittlesey** who was lost at sea.

Eastlawn Cemetery, close to Thompson Chapel, is where the first North Adams man killed in the Korean War, **Gordon A. Greene**, is buried. Just nearby, **Ruben Shay**, a veteran of three wars, and **George Guillotte** are also buried there. (His brother **Robert Guillotte** is buried in Hampton, Virginia.)

John Hitchcock of Williamstown, outdoorsman to the end, chose not to be laid to rest in a cemetery. His spirit can surely be felt on local trails or at any number of ski mountains from the Berkshires to the Rockies.

Berkshire patriots spread their influence far and wide. In the Washington area, visitors to Arlington National Cemetery should look for **Leonard and Ruth Koczela**'s resting place. Readers seeking more about her codebreaking can explore the Code Girls HQ at the US Naval Communications Annex in Washington, DC.

For further military ingenuity, visitors to San Diego can see **Admiral McCann**'s McCann Chamber at the US Naval Undersea Museum. Those traveling through Maryland can learn more about **Michael Scarpitto** at the Ritchie Boy Museum in Cascade; those in North Carolina can see a commemorative portrait of **Ishmael Titus** in Charlotte. To honor **Marc Jaffe**, you could visit the 1st Marine Division monument on Peleliu Island—or just look for him in Williamstown and thank him for his service.

The simple act of recounting these places brings back deep memories for me, of the hours spent pondering their sacrifices, of lives cut short, lives unlived. Even for those who came back, the arc of their lives was changed forever. Their honor remains.

Sources

FOR THE PAST MANY months, I have spent countless hours in the company of Berkshire patriots.

Many of my moments in this accelerated age have been spent researching information on newspapers.com, Googling, or checking various other internet sources. By far, however, my greatest satisfaction came from reading books about the patriots' lives, books brimming with history that helped me better understand and appreciate the eras and cultures they grew up in.

I wanted to share a selection with you and have attached a list in the following appendix. Many of my "Berkshire Patriots" are mentioned directly in these books. The stories are fascinating. Robert Guillotte's desperate attempt to escape from the Gestapo after his plane is shot down is heavily covered in *The History of the Hell Hawks* by Charles R. Johnson. Charles Whittlesey, Williamstown's Medal of Honor Winner in World War I, is best read about in Richard Slotkin's *Lost Battalions: The Great War and the Crisis of American Nationality.*

Ruth Koczela is mentioned in Liza Mundy's outstanding book, *Code Girls*. Daniel Petithory, a Green Beret hero, is part of Eric Blehm's *The Only Thing Worth Dying For: How Eleven Green Berets Fought for a New Afghanistan*, a must-read. Marc Jaffe, a centenarian living in Williamstown, who as a 1st Lieutenant survived the devastating assault on Okinawa, is quoted in *The Battle of Okinawa: The Blood and The Bomb* by George Feifer.

SOG: The Secret Wars of America's Commandos in Vietnam was difficult to put down. John L. Plaster's writing gives one a deeper idea of the challenges faced by Ruben Shay, the three-war veteran who lived in Williamstown. Two hundred years earlier, the rally of

Colonel Joab Stafford and his group of militia is covered in numerous books, particularly how their march from what is now Cheshire to the Battle of Bennington had a decisive effect on the American Revolution.

Happily, the books on the list, whether used for research or reading pleasure, were all transportive experiences. I hope some of you so enjoyed the brief vignettes in *Berkshire Patriots: Stories of Sacrifice* that you will be intrigued and compelled to delve further into the lives and times of our local heroes. The wanderings will be worth it. — DP

A Reading List
Selection of Source Material

John E. Atwood

Murray, Stuart. *A Time of War: A Northern Chronicle of the Civil War.* Berkshire House Publishers, 2001.

Sterling S. Burnette

Dominique, Dean, Major & Colonel James Hayes. *One Hell of a War: Patton's 317th Infantry Regiment in WWII.* Wounded Warrior Publications, 2014.

Murrell, Robert T. *317th Infantry Regiment History WWII. E.T.O. 80th "Blue Ridge" Infantry Division,* 2015.

Henry F. Caron

Ellis, John. *Eye-Deep in Hell: Trench Warfare in World War I.* The John Hopkins University Press, 1989.

Hinson, Dae. *As I Saw It in the Trenches: Memoir of a Doughboy in World War I.* McFarland & Company, 2015.

Kreisler, Fritz. *Four Weeks in the Trenches: The War Story of a Violinist.* Palala Press, 2015.

Michael F. Cirullo

Lyman, William J. *Curlew History: First Battalion.* The Orange Printshop, 1948.

George L. Curran, M.D.

Mayhew, Emily. *Wounded: A New History of the Western Front in World War I.* Oxford University Press, 2014.

Helling, Thomas MD. *The Great War and the Birth of Modern Medicine: A History.* Pegasus Books, 2022.

Scotland, Thomas & Heys, Steven. *War Surgery 1914–1918.* Helion & Company, 2012.

Peter W. Foote

Arthurs, Ted G. *Land with No Sun: A Year in Vietnam with the 173rd Airborne.* Stackpole Books, 2006.

Murphy, Edward. *Dak To: America's Sky Soldiers in South Vietnam's Central Highlands.* Presidio Press, 2007.

Thompson, Robert J. III. *Clear, Hold, and Destroy: Pacification in Phu Yen and the American War in Vietnam.* University of Oklahoma Press, 2021.

Gordon A. Greene

Blair, Clay. *The Forgotten War, America in Korea 1950–1953.* Times Books, 1987.

Bowers, William T., et al. *Black Soldier/White Army.* US Army Center of Military History, 1996.

Goulden, Joseph, C. *Korea: The Untold Story of the War.* Times Books, 1982.

Houston, Ivan J. *Black Warriors: The Buffalo Soldiers of World War II.* iUniverse, Inc., 2009.

Rishell, Lyle. *With a Black Platoon in Combat: A Year in Korea.* Texas A & M Press, 1993.

Morrow, Curtis James. *What's a Commie Ever Done to Black People?* McFarland & Company, 1997.

George L. Guillotte, Robert J. Guillotte

Dorr, Robert F. & Jones, Thomas D. *Hell Hawks! The Untold Story of the American Fliers Who Savaged Hitler's Wehrmacht.* Zenith Press, 2008.

Johnson, Charles R. *The History of the Hell Hawks*. Southcoast Typesetting, 1975.

John C. Hitchcock

Daly, Hugh Lt. *42nd "Rainbow" Infantry Division: A Combat History of World War II*. Hassell Street Press, 2021.

Hansult, William George Jr. *The Final Battle: An Untold Story of WWII's Forty Second Rainbow Division.* Mill City Press, 2018.

Marcus H. Jaffe

Feifer, George. *The Battle of Okinawa: The Blood and the Bomb*. Lyons Press, 2020.

Hunt, George P. *Coral Comes High*. Arcadia Press, 1946.

Sledge, E.B. *With the Old Breed: At Peleliu and Okinawa*. Ballantine Books, 1981.

Ruth E. Koczela

Mundy, Liza. *Code Girls: The Untold Story of the American Women Code Breakers of World War II*. Hachette Books, 2017.

Rudy Konieczny

Isserman, Maurice. *The Winter Army: The World War II Odyssey of the 10th Mountain Division, America's Elite Alpine Warriors*. Mariner Books, 2019.

Rottman, Gordon. *US 10th Mountain Division in World War II*. Osprey Publishing, 2012.

Sanders, Charles J. *The Boys of Winter: Life and Death in the U.S. Ski Troops During the Second World War*. University Press of Colorado, 2005.

Allan R. McCann

LaVO, Carl. *Pushing the Limits: The Remarkable Life and Times of Vice Admiral Allan Rockwell McCann, USN*. Naval Institute Press, 2013.

Elisha Nims

Champney, Wendy. *The Forgotten Ledge of Fort Massachusetts*. 2016.

Egleston, N.H. *An Old Fort and What Became of It* and *Norton's Redeemed Captive 1746*. North Adams Historical Society, 1991.

French and Indian War: A History from Beginning to End. Hourly History, 2017.

Haefeli, Evan & Sweeney, Kevin. *Captors and Captives: The 1704 French and Indian Raid on Deerfield.* University of Massachusetts Press, 2003.

Norton, John Reverend. *A Narrative of the Capture and Burning of Fort Massachusetts by the French and Indians, in the Time of War of 1744–1749 . . . . Written at the Time by One of the Captives.* Scholar Select, 1746.

Smith, Mary P. *Boys of the Border.* John W. Haigis, Jr., 1954.

Daniel H. Petithory

Blehm, Eric. *The Only Thing Worth Dying For: How Eleven Green Berets Fought for a New Afghanistan*. Harper, 2010.

Pepin, Rebecca. *Faces of Freedom: Profiles of America's Fallen Heroes, Iraq and Afghanistan.* Wentworth Printing Corp., 2007.

Moore, Robin. *Task Force Dagger: The Hunt for Bin Laden—On the Ground with Special Operations Forces in the War on Terror*. Random House, 2003.

John H. Quinn

Astor, Gerald. *The Mighty Eighth: The Air War in Europe as Told by the Men who Fought it.* Caliber, 1997.

Harbour, Ken and Harris, Peter. T*he 351st Bomb Group in W.W. II.* Byron Kennedy and Company, 1980.

Miller, Donald L. *Masters of the Air: America's Bomber Boys Who Fought the Air War Against Nazi Germany.* Simon & Schuster, 2006.

Stevens, Larry. *It Only Takes One: Memoirs of a Tail Gunner*. Larry Stevens, 2013.

Michael Scarpitto

Eddy, Beverley Driver. *Ritchie Boy Secrets: How a Force of Immigrants and Refugees Helped Win World War II*. Stackpole Books, 2021.

Henderson, Bruce. *Sons and Soldiers: The Untold Story of the Jews who Escaped the Nazis and Returned with the U.S. Army to Fight Hitler*. William Morrow, 2017.

Ruben W. Shay

Kelly, Francis, J. Colonel. *Vietnam Studies: U.S. Army Special Forces 1961–1971*. Department of the Army, 1972.

Plaster, John L. *SOG: The Secret Wars of America's Commandos in Vietnam*. Caliber, 1997.

Joab Stafford

Beach, David A. *Meaning of the Battle of Bennington*. Forgotten Books, 2018.

Comtois, Pierre. "Battle of Bennington: Home Win for the New England Patriots." *Military History Magazine*, 2005.

Gabriel, Michael P. *The Battle of Bennington: Soldiers & Civilians*. History Press, 2012.

Ketchum, Richard M. *Saratoga: Turning Point of America's Revolutionary War*. Henry Holt & Company, 1997.

Taylor, Maureen. *The Last Muster: Images of the Revolutionary War Generation*. Kent State University Press, 2010.

Author Unknown, *The Battle of Bennington, Second edition*, Published by SumoPress.

Ishmael Titus

Dameron, J. David. *King's Mountain: The Defeat of the Loyalists—October 7, 1780*. DaCapo Press, 2003.

Draper, Lyman C. *King's Mountain and Its Heroes.* Alpha Editions 2019.

Nell, William Cooper; Introduction by Harriet Beecher Stowe. *The Colored Patriots of the American Revolution.* Robert F. Wallcut, 1855.

Lau, R. Greg. *The Young Wagoner: Surviving Braddock's Defeat, 1755.* Newman Springs Publishing, 2020.

Maass, John R. *The Battle of Guilford Courthouse: A Most Desperate Engagement.* History Press, 2020.

Syfert, Scott. *Eminent Charlotteans: Twelve Historical Profiles from North Carolina's Queen City.* McFarland & Company, 2018.

John Welch

Hearn, Chester G. *Mobile Bay and the Mobile Campaign: The Last Great Battles of the Civil War.* McFarland & Company, 1993.

Parker, Foxhall A. *The Battle of Mobile Bay.* Forgotten Books, 2015.

Wakefield, John F. *Incidents in the American Civil War: Battle of Mobile Bay, 1864.* Book 45, Honors Press, 2000.

Reuben A. Whipple

Ferrer, Ada. *Cuba: An American History*. Scribner, 2021.

Hicks, Herbert O., Fred A. Simmons. *Company M, and Adams in the War with Spain*. Press of the Adams Freeman, 1899.

Charles W. Whittlesey

Slotkin, Richard. *Lost Battalions: The Great War and the Crisis of American Nationality*. Henry Holt and Company, 2005.

Ephraim Williams Jr.

Jenkins, MacGregor. *Sons of Ephraim and the Spirit of Williams College*. Houghton Mifflin Company, 1934.

Pew, William A. *Colonel Ephraim Williams, An Appreciation.* Scholar Select, 1919.

Perry, Arthur Latham. *Origins in Williamstown*. First Rate Publishers, 2022 (1894).

Wright, Wyllis E. *Colonel Ephraim Williams: A Documentary Life*. Berkshire Historical Society, 1970.

About the Author

Pregent (left) in 2022, visiting with Marc Jaffe (center) in his Williamstown home with former Veterans Agent Mike Kennedy (right). *[Courtesy of the author]*

DENNIS G. PREGENT, the author of *The Boys of St. Joe's '65 in the Vietnam War* and *Born in the Berkshires,* is a native son of Massachusetts. Born in Pittsfield and raised in North Adams and Adams, Pregent attended local schools, and in his senior year of high school, at seventeen, he enlisted in the Marines.

After serving almost seven years of active duty with two tours in Vietnam, he returned to the Berkshires, graduated from North Adams State College in 1975 (currently Massachusetts College of Liberal Arts) and went on to receive his MBA from the University of Massachusetts in 1977 before embarking on a lifelong career in human resources.

For over thirty-five years, until his retirement, Pregent served in HR as an international vice-president for the Evenflo/Spalding and ConAgra Food companies. His responsibilities were as far-ranging as domestic and foreign labor negotiations to international executive staffing.

Several years after his retirement in 2012, Pregent began researching and writing books focused on the beautiful and history-rich region of the Berkshires. *Berkshire Patriots: Stories of Sacrifice* is his third book.

He and his wife, Carol, have six children and fifteen grandchildren, and reside in Garner, North Carolina.

For further information or to contact him, please visit ***dennispregent.com.***